stinging trees & wait-a-whiles

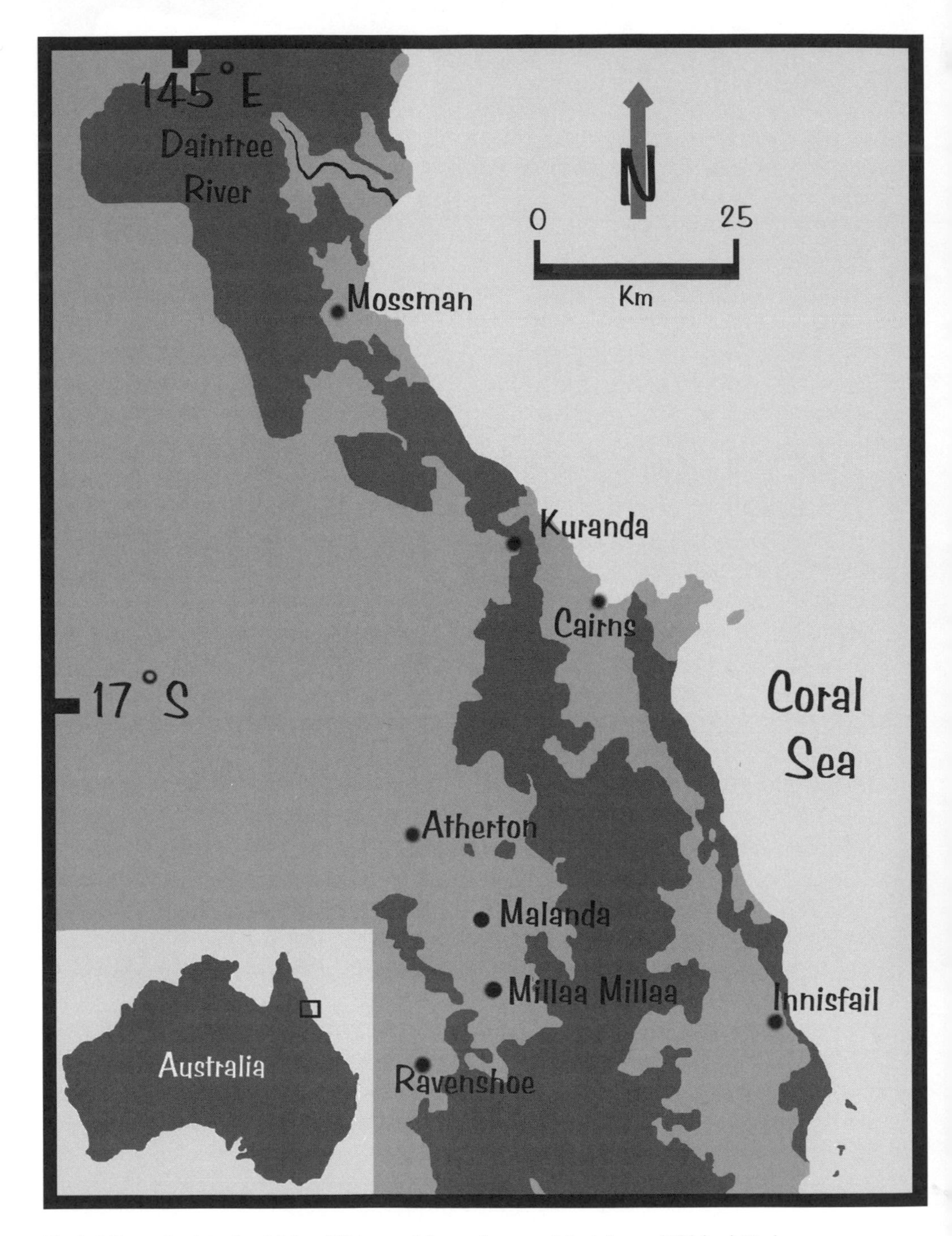

Tropical Queensland, in the vicinity of Cairns and the rural towns of the Atherton Tableland. Dark-gray areas are rainforest, while light-gray areas are deforested, or support dry forests. The Atherton Tableland stretches from Ravenshoe northward to Atherton and beyond.

stinging trees & wait-a-whiles

CONFESSIONS OF A RAINFOREST BIOLOGIST

William Laurance

THE UNIVERSITY OF CHICAGO PRESS
CHICAGO AND LONDON

WILLIAM LAURANCE is a research scientist for the Smithsonian Institution Biological Dynamics of Forest Fragments Project in Manaus, Brazil. He was coeditor of *Tropical Forest Remnants* (1997), also published by the University of Chicago Press.

The University of Chicago Press, Chicago 60637
The University of Chicago Press, Ltd., London

Printed in the United States of America

09 08 07 06 05 04 03 02 01 00   1 2 3 4 5
ISBN: 0-226-46896-8 (cloth)

Library of Congress Cataloging-in-Publication Data

Laurance, William F.
Stinging trees and wait-a-whiles : confessions of a rainforest biologist / William Laurance.
p. cm.
Includes bibliographical references (p.).
ISBN 0-226-46896-8 (alk. paper)
1. Ecology—Australia—Millaa Millaa (Qld.)—Field work—Anecdotes. 2. Millaa Millaa (Qld.) 3. Laurance, William F. I. Title.
QH197 .L38 2000
577.34′09943′6—DC21   00-020543

∞ The paper used in this publication meets the minimum requirements of the American National Standard for Information Sciences—Permanence of Paper for Printed Library Materials, ANSI Z39.48-1992.

To my Mother and Father,
who enthusiastically endorsed my pursuits in biology,
no matter how wacky they seemed at the time,

AND

To the Rat Patrol and Higher-Mammal Crew

CONTENTS

Preface *ix*

1. Assault from Above *3*
2. Stinging Trees and Wait-a-Whiles *17*
3. Friendly Leeches *31*
4. The Mystery of the Day-Glo Rats *45*
5. Invasion of Millaa Millaa *53*
6. Rainforest Politics *63*
7. Gone Troppo *73*
8. The Higher-Mammal Crew *81*
9. Something's in the Bathtub *89*
10. Old Harry *97*
11. The Wet Season *105*
12. Jenny *115*
13. The World Heritage Controversy *123*
14. Joh Starts His Run *131*
15. Right in the Thick of It *137*
16. The Rat Trials *143*

17. Guerilla Warfare 151
18. New Guinea 159
19. Missionaries, Mercenaries, and Misfits 165
20. The Togobas 171
21. Return to Oz 179
Postscript 185
Rainforest Conservation Organizations 187
Glossary 189
Species Mentioned in the Text 191
Suggested Readings 195

PREFACE

This is a very personal tale of eighteen unforgettable months I spent in the rainforests of northern Queensland, Australia, doing field research for my Ph.D. in zoology. This is not a typical account about scientific research—at least I hope not, for my colleagues' sake. Rather, it is a story about the joys and agonies of fieldwork, about zany characters and a wild clash of cultures, and about a bitter conservation battle that erupted in the midst of the whole thing. It is also a story about tropical rainforests—their remarkable denizens and the forces that threaten them.

This account describes a time of intense challenges and growth for me. I began my field project as a rather impetuous young man, and ended with a deeper appreciation of many things. In writing this book I've tried to recount events as they actually happened, resisting the temptation—and indeed the suggestions of a few—to filter out the more rambunctious and raw parts of the story. So here it is—warts and all. A friend once told me that writing a book is like hanging a sign around your neck that says "Kick me!" I just hope my derriere is up to it.

I sometimes struggled in writing about the events of a dozen years ago. The vagaries of time clouded my recollection. It also was difficult to verify certain details—especially for secondhand stories, which I have nonetheless done my best to relate—given that I live half a world away from Australia today. I am greatly indebted to Kate Lehrer, Christie Henry, Carol Saller, Emilio Bruna, Kate Uehling, Chris Funk, Steve Comport, Carol Hautót, John Winter, Claire Laurance, Susan Laurance, and five anonymous reviewers for countless suggestions to improve the manuscript. In addition, Nicola Moore, Daryn Storch, Aila Keto, Ingrid Nielsen, Charles Baker,

Christine Buehler, Steve Comport, Tony Irvine, James Patton, Aaron Bauer, and Bill Lidicker kindly dug up piles of invaluable information, while Alexandre Kirovsky created the maps of northern Queensland and New Guinea. Of course, any errors of fact or interpretation are entirely my own.

Two people deserve special credit. My wife, Susan, tolerated my night-owl habits and provided frequent advice on the text, including vital insights into the Australian psyche and politics. Equally phenomenal was my editor, Christie Henry, who coached and cajoled me every step of the way. Christie and I had worked together before while producing the book *Tropical Forest Remnants: Ecology, Management, and Conservation of Fragmented Communities* (edited by myself and Richard Bierregaard, University of Chicago Press, 1997), so I knew she was a wonderful editor. But her commitment, constructive criticisms, and editorial skills were truly without equal.

stinging trees & wait-a-whiles

assault from above

The illuminated raindrops were like sparkling crystals. They plummeted downward by the millions, every drop seeming to converge at my face. The effect was mesmerizing, almost hypnotic—like flying at light speed through a galaxy dense with stars.

I paused for perhaps the twentieth time in the last hour to wipe the rain from my glasses. I usually wore contact lenses but tonight I'd been lazy, opting for glasses instead. I was glad I had. The rain would have been intolerable otherwise, pelting my upturned face like a kind of self-imposed Chinese water torture.

I kept sweeping the vegetation: up, down, up, down, methodically covering each tree from different angles with my handheld spotlight. The powerful light was like a giant paintbrush, and I tried to paint every tree, every vine, every epiphyte as completely as possible.

The track was sodden and muddy, and I slipped and staggered yet again, nearly careening into a thorny rattan. The damned red mud. Just add water, and the rich basaltic soil was transformed into a sticky quagmire, clinging in great congealed globs to boots and truck tires.

The hardest part of this whole exercise was trying to hike along quietly over steep, rugged terrain while continually looking upward, my eyes and ears focused on the forest's dense canopy above. The back of my neck was sore, though marginally less so than the last couple of nights. My muscles were still adjusting to the mistreatment I was doling out each night, and my shoulders ached from the big motorcycle battery I carried in my backpack to power the spotlight.

As I stumbled along, the forest was far from silent. Accompanying the whisper of falling rain and the squish of footsteps was the rhythmic drone of cicadas, their metallic calls reverberating inside my ears. Occasionally a mysterious sharp bark, or a gurgling rasp, emanated from deep in the forest. Tonight the crook car-horns—a rainforest frog—were also calling intermittently, their insistent nasal honks verging on the humorous.

While gazing upward my leg bumped into something hard. A massive tree had collapsed across the logging track. It must have fallen recently, as the rattans and thick lianas that snaked along its trunk were still alive. I threw one leg over the trunk and suddenly there was a thunderous crashing above me. Something big was plummeting down from high in the canopy. I tried to swing my light around but I was awkwardly straddling the fallen tree. The falling object landed with a powerful thump only yards away, then sprang to life and bounded frantically into the dense undergrowth. I pivoted wildly to see it but, top-heavy from the big battery, slipped on the log and fell flat on my back.

My light went out and I was enveloped in total, suffocating blackness. My heart pounded in sync with the creature's distinctive thumping leaps that echoed through the forest.

Jesus! I gulped a few breaths, then slowly wrestled myself from the mud's grasp. Though the back of my neck still tingled from fear, I realized I had finally encountered the beast. Only one creature in the Australian rainforest—probably only one animal in the entire world—behaves like that, crashing thunderously down from the treetops then loudly bounding away. I pulled out a notebook and flashlight and shakily began recording my observation: 11:04 P.M., 1 Lumholtz's tree-kangaroo.

❧

I had arrived in Australia just two weeks earlier, in June 1986. The customs officer at Cairns International Airport eyed me curiously as I struggled to push three overloaded trolleys up to his inspection booth.

"You traveling alone?" he asked, the lifting inflection at the end of his question revealing a kind of incredulity. The man's no-nonsense attitude and carefully pressed, pale-blue uniform demonstrated serious professionalism.

"Yes," I replied breathlessly. "This is all my stuff. I'm here to do a research project in the rainforest, probably stay at least a year."

Under furrowed brow, the officer examined my research visa carefully. I got the impression he hadn't seen many of those. After a long pause he said, "So what have we got here, exactly?"

This was a critical moment. Piled onto my trolleys were three big wooden crates, four extra-large canvas duffel bags, an army-surplus tote bag, two old suitcases, a camera bag, and a briefcase. If he decided to search my gear—which was fully within his rights—I could be stuck here for hours. The thought daunted me, as I was already dog-tired from twenty hours of flying, not to mention frenetic months of planning and preparation for this trip.

I gathered myself and launched into an energetic explanation of each bag and crate, exhaustively detailing its contents, pausing only to snatch quick breaths. My body language, my strong eye contact, and my prideful countenance all indicated a keen desire to describe my great pile of baggage in the most painstaking detail. For some arcane pieces of research equipment, I even provided a little explanation of what each was to be used for.

I was actually starting to enjoy myself when the officer's eyes began to glaze with boredom, and I knew I'd prevailed. "Uh, that'll be okay, mate. Go ahead on through." As I left his booth I even managed to adopt a slightly disappointed look, as though I still had many interesting things to tell him.

Through the automatic doors of the airport the heat and humidity waited like a solid wave. At 17 degrees south latitude, Cairns is squarely in the tropics. A city of fifty thousand residents augmented by ten or twenty thousand tourists and backpackers, Cairns is popularly known as the Gateway to the Great Barrier Reef—that band of coral bommies, reefs, and atolls that stretches some twelve hundred miles along the northeastern coast of Australia.

I began sweating instantly. Grubby and tired from my long flight, I wanted nothing more than a cool shower and at least ten hours of sleep. But I couldn't do that yet. I had to stay awake at least until nightfall, eight hours away. Otherwise I'd need a week to recover from jet lag and to synchronize my biological clock with eastern Australia.

I also had two important things to do today. I waved to a taxi-van driver and explained where I needed to go. We hefted my gear into his van, then drove slowly out of the airport, crossing the remnants of tidal marshes and tall mangrove forests that had survived the draining and clearing of airport construction. The mangroves, inundated by the tides twice a day, either had spiderlike roots that perched above the dense mud, or dozens of knobby protuberances—called pneumatophores—that poked up like warty little gnomes to allow their roots to breathe.

We headed southward toward the city center of Cairns—pronounced "Cannes" by the Aussies, like the film festival. I was stunned by the proliferation of new hotels, restaurants, and other buildings since I'd visited only two years before. The van driver confirmed that Cairns was booming, largely because of the influx of tourists arriving via the new international airport. Most came to dive or snorkel on the reef. With its many hostels and cheap scuba-diving lessons, Cairns was especially popular among young, nature-loving travelers from Europe, the United States, and elsewhere in Australia.

We veered west toward the city's commercial area and I soon saw what I was looking for: used-car dealers. During the next two hours we visited a half dozen car lots, bantering and haggling with the dealers. Eventually I found a snub-nosed Mitsubishi pickup, six years old, with double rear tires. It was in decent shape and the price was right—the equivalent of three thousand U.S. dollars. In a short time I was its proud new owner. I paid the van driver, heaved my gear in the back, then drove off, northward, toward my second destination.

I drove cautiously. Like the English, the Aussies drive on the wrong side of the road—the left. Everything is reversed. The steering wheel is on the right and none of the switches are where they're supposed to be. I kept turning on the windshield wipers when I tried to use the turn signal, and the gears ground painfully as I spastically tried shifting with my left hand. On top of all that I had to focus on not making any horrific driving errors. When I'd first tried driving in Australia—in Sydney, of all places—I'd only narrowly avoided disaster. I'd kept veering to the left, especially during left-hand turns, and running up onto the curb. Pedestrians on street corners literally had to dive out of my way and had screamed great streams of oaths at me. I was determined not to have a repeat performance now.

I began to relax a bit as I left the city limits of Cairns. I crossed the Mulgrave River with its "Beware of Crocodiles" sign prominently displayed along the bank to ward off swimmers. The sign wasn't a joke. Two years before, I'd swum across the wide mouth of the Mulgrave and back. Later I learned how dumb I'd been. Six people were killed by crocodiles that year in Australia, including a young woman taken while skinny-dipping in the Daintree River, only a half hour north of here.

Paralleling the coast, I followed the state highway through miles of tall sugarcane fields, the grasslike stems rippling in the breeze. Inland, to my left, the volcanic peaks of the Great Dividing Range reared up dramatically from the coastal lowlands, their angular escarpments blanketed in dense rainforest.

On the right side of the highway I spied what I was looking for—a small sign with the letters RSPCA painted on it. I turned down a bumpy gravel road and finally reached a small parking lot and office. Adjoining the office were several fenced runways and a sign: The Royal Society for the Prevention of Cruelty to Animals. I had come to the local RSPCA—Cairns's animal shelter—to get myself a dog.

I'd been wanting a dog for a long time, but my mobile and rather hand-to-mouth existence had prevented it. During the past three years I'd been a graduate student at the University of California at Berkeley, living in the San Francisco Bay area along with seven million other inhabitants. Cramped into small apartments or shared student houses, any pet more demanding of space than a goldfish was simply out of the question.

Now, I felt that I needed a companion. Only a few months before I left for Australia, my girlfriend, Beth, had broken off our relationship. We'd been a serious couple, living together with plans for marriage, and I was devastated when she suddenly decided to end it. Until then, I'd been planning for her to join me in Australia, just as she had two years earlier.

I couldn't really blame her. Like a lot of graduate students, I'd been utterly caught up in the intellectual whirlwind at UC Berkeley. The demands on grad students were daunting. On top of a twenty-five-hour-a-week teaching job, I'd been taking several courses and graduate seminars every semester, studying for my foreign-language exam, and cramming for my orals—the much-feared, four-hour verbal examination that determined whether a graduate student was qualified to pursue a Ph.D. I'd also been frenetically writing grant applications to fund my research in Australia. One of the Berkeley faculty liked to quip that "a stressed graduate student is a good graduate student," and at times I wondered if this wasn't really the department of zoology's unofficial motto.

To get through the semesters I often worked until midnight on weeknights, and eight hours on Saturdays. In a moment of weakness, I'd begun smoking cigarettes as a buffer against the stress. Beth was infuriated by my smoking—in her view, smokers were inferior life-forms—and after tolerating my nutty schedule for two years, she'd finally had enough.

In the few months since the breakup, I'd felt like a zombie. A numbness gripped me, and I drifted through each day feeling sad or not much at all. As the date of my departure approached, I'd grown increasingly happy with the prospect of leaving Berkeley with its frantic pace and raw, painful memories.

The one thing that had kept me going was the prospect of a big stint of fieldwork in the rainforest. I'd long been intrigued by the mystique and complexity of tropical rainforests, and by the daunting challenges of studying them. Equally important to me was that they were being destroyed at an astonishing pace. I'd gone to great pains to design a research project that had real-life implications for forest conservation, and I was busting to get started.

I'd gotten a final jolt the day before my flight. I walked through the door of my office and there was Beth, looking anxious and laden with bright flowers. She'd come to wish me luck. For a few moments I was speechless. Mercifully, she sensed my discomfort and soon left, a sad smile and kiss her final legacy. I was so busy that day I barely had time to think about her, but during the long, quiet flight from San Francisco to Australia, chasing the setting sun across the Pacific Ocean, my mind had swum with memories and thoughts of what might have been.

He had sad, sensitive brown eyes. He was also thin, his fawny coat sporting brown spots and a few odd, grayish splotches. The kind lady at the RSPCA said he was about six months old, an Aussie cattle dog—a kelpie–blue heeler cross—and that he'd been mistreated by his former owner. I said I'd take him, and paid the kennel fee. He seemed pleased but nervous as I carried him out to my truck, his tail wagging faintly.

He cowered quietly in the front seat. As we continued northward I raced into a store to buy a collar, tins of dog food, and a package of doggy tidbits. We then veered westward and began ascending the steeply curving road that weaves into the rainforest-clad mountains north of Cairns.

My new friend needed a name. I pondered the possibilities as we drove along, the air cooling noticeably as we left the lowlands behind. As we climbed, the rainforest became slightly shorter in stature. Trees were festooned with bird's-nest ferns and orchids, or draped lianas and ropelike rattans down their trunks. Patches of bougainvillea flashed by, the sweet scent of their orange flowers filling the air.

I glanced at a road sign showing distances to nearby towns. Suddenly I realized I'd found a great name for my dog—not only Australia's wettest town but a great Irish name to boot.

I began feeding him tidbits one by one, repeating his new name over and over again. Slowly he was gaining confidence, wagging his tail and edging closer to me. Then he suddenly decided to climb into my lap, catching me off guard with a slurping lick in the face. The truck weaved erratically as I grimaced and tried to wipe the saliva from my mouth. By the time we'd topped the mountain my dog was responding to his new name—Tully—as if he'd been born with it.

Tully and I passed through several miles of upland rainforest, which became stunted and sparse as we approached the leeward side of the mountain. Australia's tropical rainforests are very limited in extent, confined to a narrow band skirting the northeastern coast of the continent. Nowhere is the band more than thirty miles wide, and here, at one of its narrowest points, it barely spans ten miles.

Passing by the mountain village of Kuranda—a quaint, artsy town known for its Saturday tourist markets—the rainforest changed abruptly to open woodland. Stately rainforest trees were replaced by gums, wattles and she-oaks, with a ground layer of kangaroo grass punctuated by large, oddly shaped termite mounds. Proteas and banksias, their flowers shaped like bottle-brushes, grew in dense clumps along a few meandering creeks. In contrast to the verdant lushness of the rainforest, the colors here were muted pastels—faint blues, greens, and pinks. You could learn to love the eucalypt woodland with its bleak, austere beauty, but it took a little time.

We'd reached an upland plateau. The afternoon sun was lengthening, so I sped up. Tully stuck his nose out the window, his eyes squinting and ears flapping wildly in the breeze, while I tuned the radio to a country station. It felt good. I was tired but gradually relaxing as we sped toward our destination. The radio was playing *Hey True Blue*, a classic ballad of Australian mateship, the song's theme that you should always stand by your mates—especially when they were in trouble. As I would eventually learn, some gutsy people in these parts took such lyrics to heart. But I was blissfully unaware of the strife and turmoil that lay ahead. Had I ever imagined what I was getting myself into, I'd have been tempted to just spin the truck around and run for home.

In the tropics, the sun plummets below the horizon with unsettling speed. We'd turned southward an hour earlier and were now approaching the town of Atherton, namesake of the Atherton Tableland. Climbing steadily, we'd witnessed another transformation as the landscape became wetter and greener again, the first patch of rainforest appearing just before we entered the town.

We drove through Atherton in dwindling light. Tully lay curled on the

Unadorned rock wallabies live on scattered rocky outcrops in dry woodlands of northeastern Queensland. This youngster was hoping I would share my lunch with him. PHOTO BY AUTHOR.

seat, his nose tucked under his tail. The temperature was dropping. Leaving the town, we turned southward again, picking up speed as the last pink vestiges of the day faded from the horizon. It began raining, slowly at first then in greater intensity until it nearly defeated the frantic efforts of the windshield wipers. Patches of pallid fog hung in gullies, reducing visibility to a few dozen yards. The wet pavement of the undulating road seemed to suck up my headlights like a black hole.

After a very long hour, we neared our destination. The rough weather had drained the last of my enthusiasm; now I only wanted sleep. We left the road for a gravel track, crossed a rickety bridge, then ascended a hill on top of which stood a small, dilapidated white house that had never looked so appealing. The porch light was burning, and as we drove up the door flew open and a one-legged man with a great head of unruly hair and long beard came hopping out, grinning hugely. It was my old friend Geoff Downey, and he informed me I was just in time for tea.

In Australia, tea doesn't always mean tea. It often means dinner, too, and I hoped that was what Geoff meant right now. I'd learned about teas and dinners the first time I met Geoff back in '84, on my very first day on the Atherton Tableland. We'd struck up a conversation while I was looking at a small patch of rainforest that abutted Geoff's dairy farm, and he and his wife, Jo Ann, had invited Beth and me around for tea that evening. We showed up at the appointed time expecting nothing more than a few cups of tea and perhaps a little cake to go with it. Instead, we were regaled with a fine country dinner, which turned into a very long, drunken evening, which then evolved into an overnight stay on Geoff's sofa bed. Geoff and Jo just wouldn't let us leave, their hospitality boundless. We ended up camping in their living room for several weeks, and would've stayed longer if Beth hadn't finally demanded that we shift to the local caravan park. Geoff was sincerely disappointed when we left, and I was a little sad too.

Geoff is one of those rare people who, despite exceptional adversity, doggedly refuses to let life drag him down. Not that it isn't a struggle. In 1981, Geoff had been driving fast down a country road when he crashed into a car that pulled out suddenly from a hidden driveway. The car's two occupants were killed, and Geoff was horribly injured in the crash, spending hours trapped bleeding in the tangled wreckage. Geoff's left leg was so badly mangled that it had to be amputated, just below the knee, and he sustained severe head injuries. For three weeks he remained in a coma from which his doctors were by no means confident he would ever recover.

Geoff had faced tremendous challenges since the crash. He had no memories of the accident itself, the details inferred by the police from skid marks and other physical evidence. He was ruled in a court of law to have been substantially responsible for the accident, and thus was unable to collect government injury compensation. His daunting medical expenses were largely covered by Australia's socialized medical system, but he had faced months of painful recuperation which had created severe financial worries for Jo Ann and him. Somehow, Geoff had held onto their small dairy farm, which, despite his disability, he ran all by himself.

It wasn't easy. I could see how tired he often was, how frustrating it must be to have one's mobility so greatly hampered. But I still hadn't seen Geoff's greatest curse. One evening we were watching a nature show on TV about the incredible world of insects. A particular scene showed a grasshopper eating a blade of grass, the dubbed-in crunching sounds realistic and loud. Geoff, who'd been relaxing in an easy chair, suddenly arched his back sharply and moaned, his jaw and eyes clenched. The color drained from his face, and he squeezed his left leg so tightly that his knuckles turned white. For minutes he barely made a sound, just sat there slowly rocking back and forth, beads of sweat growing on his forehead.

Phantom pains. I'd heard of them, of course, but had no idea how frightening, how awful they could be. Later, Geoff told me about how they appeared without warning (tonight, it was the crunching sound that brought them on); how badly they hurt, as though the missing leg was being stabbed with white-hot needles; and how no medicine could alleviate the pain.

❧

That Geoff faced his life with humor and grace and unfailing generosity is, in my view, one of the great triumphs of the human spirit. I liked Geoff, and respected him, and it was for this reason that I found my protective hackles rising whenever someone was less than kind to him.

You see, Geoff was different from most folks on the Atherton Tableland—certainly different from other dairy farmers. For one thing, he liked to indulge in a little marijuana, partly to alleviate his daily rigors and partly to make the bursting phantom pains a little more tolerable. Geoff's pot-smoking habits were no big secret; indeed, there were few real secrets on the Atherton Tableland—not for long, anyway. Twice, the local police had raided Geoff's house and found small amounts of marijuana in plain view (in Queensland, no search warrant is needed if illicit drugs are suspected on the premises). On both occasions Geoff was fined nearly a thousand dollars.

In addition to smoking pot, Geoff was, well, a little eccentric. Some people said he was never quite the same after the car wreck. I hadn't known him before his accident, but I can attest that he held some rather unconventional attitudes.

For one thing, Geoff was an organic dairy farmer. He refused to use fungicides, herbicides, or any other -cides on his property. That was well and fine in theory, but in practice Geoff's cattle paddocks were burgeoning with weeds, and the neighboring farmers complained that Geoff's weeds were spreading into their clean grassy paddocks.

In addition, Geoff was a pacifist, a nature lover. He was happy to share his house with a brushtail possum, his paddocks with wallabies, and he refused to kill the red-bellied blacksnake—a venomous species—that sunned itself on the road a few dozen yards from his house. This was practically unheard of—especially the snake thing. The only good snake was a dead snake as far as most locals were concerned.

Finally, Geoff was convinced that seaweed was going to save modern agriculture. Not just sort-of convinced, or moderately convinced, but utterly, overwhelmingly, profoundly convinced. He was to seaweed what St. Peter was to Christianity, and he spread the gospel of seaweed to anyone and everyone willing to listen. Geoff used seaweed, and exclusively seaweed, to fertilize his pastures. And not just any seaweed—ground Norwegian kelp, rich in trace minerals. It was expensive, but he swore by its results.

In and of itself, the use of seaweed fertilizer was not as unusual as it might sound; what was unusual was Geoff's religious zeal in promoting it. You see, most dairy farmers on the Atherton Tableland suffered from low pasture productivity—a common occurrence in the tropics, where eons of heavy rains have leached minerals and nutrients from the soil. The government maintained minimum protein standards for milk, and the tableland's dairies were often at or below the legal limit. To bolster their low productivity, dairy farmers doused their paddocks with expensive fertilizer—nitrogen, phosphorus, potassium—the three nutrients most needed by plants in depleted soils.

Geoff's point, which might have been received more sympathetically had someone wearing a suit and tie been advocating it, was that the soils lacked more than just these three nutrients; they were also missing crucial minerals and trace elements—copper, magnesium, zinc, and the like. And what fertilizer contained huge natural concentrations of all this good stuff? Why seaweed, of course.

But something in Geoff's character seemed to threaten the conservative folk in these parts. He simply didn't conform. In Berkeley, Geoff would have passed as just another Grateful Dead fan or burnt-out Vietnam vet, hardly meriting a second look. But here on the Atherton Tableland, with his wild hair and beard and unusual attitudes, he was, to a lot of people, a kind of pariah.

❧

As I hopped up onto his porch and shook his hand, I could tell the past two years hadn't treated Geoff kindly. He looked thinner, grayer, more tired. Over a beer he told me that Jo had left him a few months earlier. He seemed to take that all right, but she'd also taken their little girl, and Geoff was very cut up about that. The dairy hadn't been faring well, and now Geoff's brother, who owned a 50 percent stake in the farm, wanted to sell out and claim the cash. You couldn't fault him, really, but Geoff's heart and soul were on that farm—it was all he had to live for.

Though I was flogged from my travels, the cold beer revived me, and we stayed up for hours catching up on local news and gossip. I made a nice bed for Tully on the porch and wrapped him up against the damp chill in my old letterman's jacket.

Delirious from exhaustion and beer, I finally collapsed at midnight on Geoff's couch. Like all dairy farmers, Geoff arose at 5 A.M., but I slept like a dead man until noon.

I needed copious coffee and a few cigarettes to get my heart started the next day. By about two in the afternoon I started reclaiming my equilibrium, and after a long hot shower decided I was feeling energetic enough to go for a drive. Geoff was off in a paddock somewhere, so I left him a note, threw Tully into the front seat of my truck, and rumbled off toward the nearby town—Millaa Millaa.

It was a lovely day, all vestiges of the previous night's rainstorm having disappeared entirely. The sky was milky blue but for a few scudding clouds. As we bumped along the serpentine road, I got my first good look at the countryside since being here two years before.

It was an extraordinary place from a biologist's perspective: part Ireland, part Costa Rica—and part Krakatoa. Like Ireland, the verdant hills supported a quiltwork of cattle pastures defined by neat hedgerows and fences. Like Costa Rica, the vegetation that persisted in scattered patches and grew along the myriad streams and rivers was dense tropical rainforest.

But like the volcanic aftermath of Krakatoa, the overwhelming impression was one of ecological destruction. The area had originally been blanketed in rainforest, but a century of European colonization had virtually denuded the landscape—200,000 acres of forest had disappeared. Today, aside from the steep mountainsides that encircle the tableland, forest survives only as scattered islands, surrounded by a hostile sea of pastures, crops, and homesteads. This wounded landscape was the reason I'd traveled eight thousand miles: to study the process of biological impoverishment that occurs when you fragment a tropical rainforest.

I'd nearly reached Millaa Millaa, but decided on a whim to veer westward, turning onto a snaking track that climbed a nearby mountain to the Millaa Millaa lookout. In a small parking lot, Tully and I strolled up to a guardrail designed to protect the unwary and the drunk from an abrupt precipice below.

We stared out at a spectacular view of the southern Atherton Tableland: a patchwork of farms, pastures, and islands of relict forest. On the eastern horizon the coastal mountains reared up like angry monoliths. Visible in the

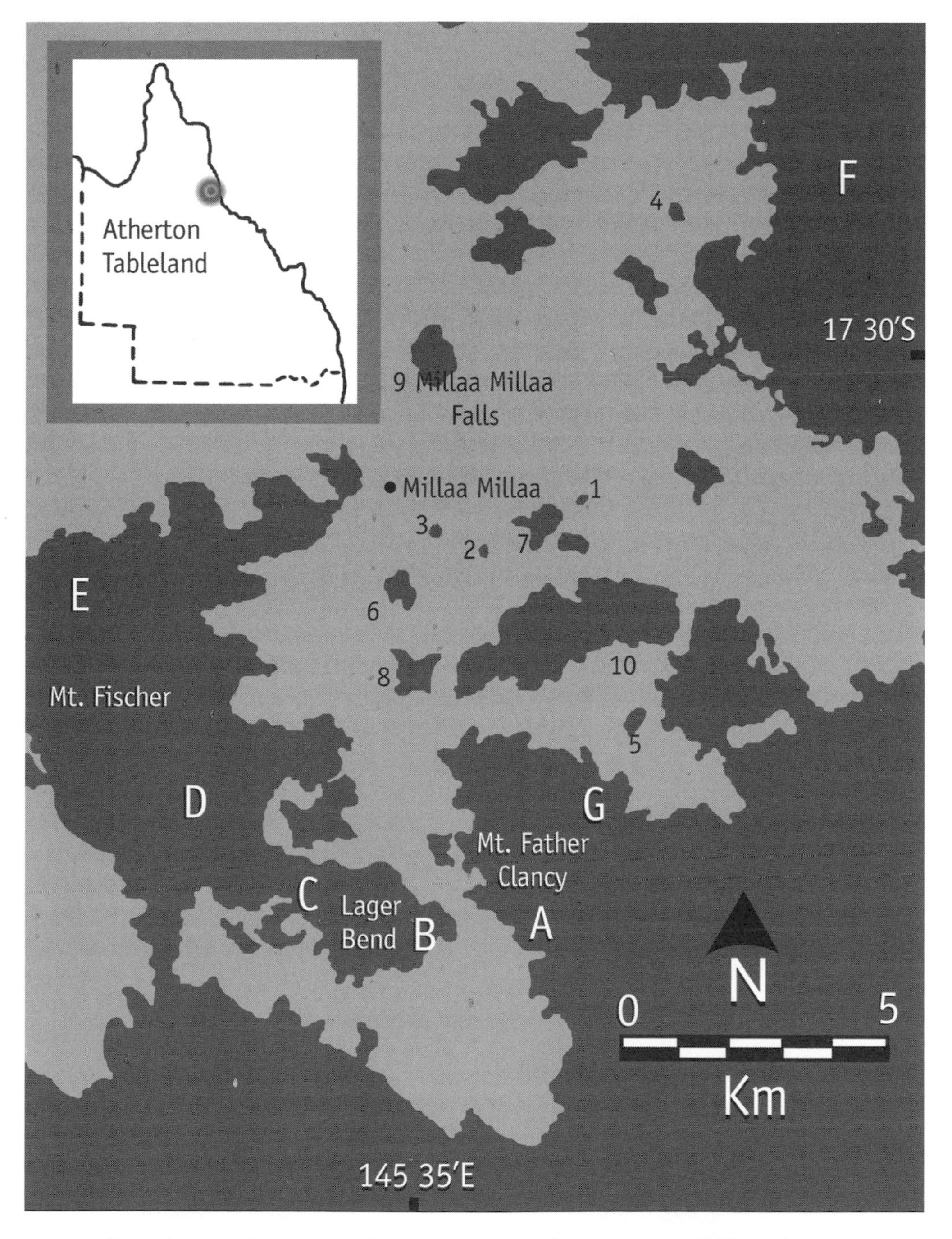

Our study area on the southern Atherton Tableland, near the tiny township of Millaa Millaa. Light-gray areas are mostly cattle pastures, while dark areas are rainforest. The forest fragments we studied ranged from about three to fifteen hundred acres in area, and are numbered from 1 to 10 in order of increasing size. "Control sites" in continuous forest are indicated by letters A-G.

distance were Queensland's tallest mountains, Bartle Frere and Bellenden Ker, their peaks wreathed in clouds. For the most part, the coastal range formed an irregular, sawtoothlike barrier between the tableland and the nearby coast. But directly east was a low saddle between the mountains. From here, on particularly clear days, you could see the shimmering blue of the Coral Sea.

The saddle was far more than an obscure geological feature. It was, in fact, one of the reasons I was so keenly interested in this particular landscape. Through the depths of time, the saddle in the mountains had profoundly influenced the rainforests of Millaa Millaa and the many species of animals and plants that lived here.

To understand the saddle's significance, one needs to step back and look at the bigger picture. When most people think of Australia, the images that spring to mind are the arid outback with its vast red deserts sparsely populated by kangaroos and wallabies, and sleepy koalas perched in eucalyptus trees.

But this is a misleading and, geologically speaking, very recent phenomenon. For most of the last sixty million years, Australia was a verdant, dripping continent, dominated by tropical and subtropical rainforests. Only in the past five million years or so have the rainforests contracted, the result of dramatic changes in the earth's climate. Eventually they retreated to the very wettest areas, especially northeastern Queensland, where prevailing southeasterly winds dump abundant rain on the coastal mountain ranges. There, the last vestiges of Australia's tropical rainforests still persist.

But the climatic upheavals had one final, capricious trick to play on the rainforests. During the great Ice Ages of the Pleistocene—the most recent ending only ten thousand years ago—the earth became much cooler and drier. Vast ice sheets spread across the higher latitudes. In the tropics, the rainforests withered and shrank into isolated pockets, persisting only in areas of exceptional humidity.

And this is when the saddle of Millaa Millaa came into play: as a window to the sea with its moist, life-giving winds. Because of the saddle the cloudy rainforest here had survived the millennia, had persisted for tens of millions of years, even during the Ice Ages. Numerous plant and animal species, many of them primitive—virtually living fossils—existed here and nowhere else on earth. The Millaa Millaa rainforest was one of Australia's most crucial biological hotspots, a stable refuge from the ravages of time.

That is, until the Europeans arrived a century ago.

stinging trees and wait-a-whiles

Millaa Millaa is a classic "one-pub town," a vignette of rural Queensland.

I drove slowly down the main street, passing a general store where one could buy anything from ladies' hats to syringes for injecting cattle. Opposite was the newsagent, purveyor of newspapers, magazines, and the like; adjoining that was a livestock-feed store. A stone's throw further was the hotel and pub—the town's social center. At the end of the street, a takeaway sold burgers and fried food.

And that was the main drag of Millaa Millaa.

A perpendicular street led to a post office and gas station. Another, paralleling the main street, held rows of tidy wooden houses, some perched on six-foot stilts to help resist marauding termites. Most houses were fronted by white picket fences overrun by vines; many had colorful rose bushes. Halfway along the street was a one-man police station, fifty yards beyond that a small grade school and then an ambulance center. A footie field—for rugby—and a nine-hole golf course rounded out the attractions of Millaa Millaa, population 320.

Oh, yes, there were also two small wooden churches. One Anglican and one Catholic. Millaa was unusual by Queensland standards for having more churches than pubs.

I called in at the Millaa Millaa Pub and over a cold beer mentioned to the bartender that I was looking to rent a house in the area. He was friendly

A local resident snoozes in front of the Millaa Millaa pub. PHOTO BY RICH ZWALLER.

enough, saying he'd pass the word along. But the pub's patrons—all men—had suddenly grown quiet when I walked in the door and were stealing curious glances at me. I'd expected as much. I'd grown up on an Oregon cattle ranch and understood the natural aloofness of country folk.

The key to getting along in a place like Millaa Millaa is to remember a few basic rules. First, don't ever act clever or sophisticated, like you're somehow better than other folks. Second, take the time to get to know people; things have their own particular pace in the country. Third, realize that most folks are hardworking, honest, and basically very straight—they say what they mean and mean what they say. Finally, always remember that if you piss somebody off, you may well turn the whole town against you.

Diplomacy, under these circumstances, is vital, and I was determined to start off here on the right foot. Most of the rainforest patches I needed to study were located on local farms, and the last thing I wanted was to get offside with anybody.

❧

I introduced myself at the takeaway, and immediately warmed to the proprietors, a couple in their fifties named Des and Helen. They'd recently moved here from Darwin in the Northern Territory and complained about how cliquish everybody was in Millaa. "It's bloody well not good enough to be born here," Des lamented; "your parents and grandparents have to hail from here too."

I ordered fish-and-chips and mentioned my research project. They were immediately interested, so I explained that I was attempting to understand which species of native mammals—everything from possums and tree-kangaroos to rodents—could survive in the area's remnant patches of rainforest. I told them we'd be working day and night—trapping and spotlighting—to census mammal populations. Des warned me to watch out for pot-growers in the rainforest. "Some of them bastards use booby traps to protect their crops and would slit your throat if they thought you'd dob 'em in to the cops," he said, his pale eyes suddenly serious.

As I left the takeaway I mentioned that I was looking for a house to rent. This really was unnecessary. I'd made my needs known at the pub and was certain that Millaa's verbal telegraph—gossip traveling at the speed of light—was already working its wonders.

❧

Within the week I'd found a suitable house, just across from the Catholic church and a hundred yards from the main street. It was a rambling, four-bedroom farmhouse—almost too big for my needs—with a green fiberboard exterior and corrugated aluminum roof. I'd rented it from a young couple with kids who'd decided to move to Cairns; work had become scarce since Millaa's sawmill had closed three years before.

The rent was quite outrageous by local standards. But I was a "rich Yank"—all Yanks were rich—so I wasn't shocked at having to pay the equivalent of three hundred bucks a month. The house was furnished: beds, couch, kitchen table, chairs, washing machine. There was a gas stove for cooking and a wood stove in an alcove off the kitchen for heating. And it had an enclosed porch, a big fenced yard for Tully, and sheds out back which would undoubtedly come in useful for something. Mango trees and flowers and vines grew in profusion.

All in all, I was very satisfied.

A small disaster struck the next day. Earlier that week I'd been organizing my gear at Geoff's place and assembling a bunch of prefabricated cage traps I'd manufactured by hand in Berkeley. I'd also driven down to Cairns to pick up Tim Thompson. Tim was a former student of mine at Berkeley who'd been intrigued by my descriptions of the Atherton Tableland. Tall and baby-faced with wavy red hair, Tim had paid his own airfare to be here and would be staying for a month to help with fieldwork.

Now that Tim and I were installed in the house, we needed an electrician. Australia uses a different power system than the United States—220 volts instead of 110—and I'd brought along quite a lot of electrical equipment: computer, printer, battery chargers, rechargeable batteries, and some small power converters. I wasn't sure if my little power converters were up to the task, and figured it'd be worth paying an electrician to get things right.

I'd asked around for an electrician and was directed to Dave, a gangly chap of about thirty, who immediately came over to the house. I explained the situation while Dave listened with a quiet, thoughtful expression which I assumed he adopted when he was pondering watts and ohms and technical stuff like that.

After several moments of seemingly deep thought he said, "Well, let's give it a go." He grabbed one of my battery chargers and attached it to a power converter, then plugged the whole thing into the wall socket. Almost instantly there was faint burning odor. Dave reached out to unplug the converter but let out a loud yelp—the converter was melting before our eyes! As he sucked on his burnt fingers a stream of acrid smoke rose up from the charger.

"Uh, maybe this isn't such a good idea," I said, slapping frantically at the melting converter to disengage it from the socket. But Dave was annoyed and determined. He grabbed my second power converter and a heavy duty charger I'd planned to use for spotlighting batteries, and quickly plugged them into the wall.

The house's lights flickered. For a tense moment everything seemed okay, but then the second converter started melting and the charger suddenly popped and was lifeless.

This was getting serious.

To hide my growing annoyance I hastily lit a cigarette and told Dave I was going to phone an electrical-goods store in Cairns. I walked into the kitchen and began thumbing through the phone book. But as I dialed a crackling sound emanated from the other room. I dashed in to find Dave staring at my printer, which gave off a brilliant flash that made us both jump back, followed by a dense cloud of black smoke.

"What the hell were you doing?!" I yelled indignantly. Then I caught myself, remembering how vital it was to get along with the locals.

As he left Dave was profusely apologetic, arguing that I didn't need to pay him. But despite all the damage I insisted he take twenty bucks; I didn't want anyone saying I didn't pay my bills. As he drove off I mentally added up the value of the equipment we'd just destroyed, wondering where I was going to get another thousand dollars to replace it.

Feeling deflated, I walked over the pub to drown my woes in a cold beer. The bartender saw I was looking glum and asked if everything was all right. As I began explaining about Dave and all the fun we'd just had playing with electricity, he started giggling, then laughing out loud, then finally roaring and pounding the bar, the tears streaming down his face. "Ohhhh," he said, wiping his eyes, "didn't anybody warn you about Dave?"

When I replied rather testily that they obviously hadn't, he guffawed, "He's the dumbest bastard in the whole damn town! He can wire a lightbulb all right, but you'd *never* let him do anything out of the ordinary!"

As I digested this the bartender explained that Dave's dream had been to join the Australian navy. But he'd flunked the entrance exam three times in a row. The last time, they'd told him he couldn't take the test again. "Now, mate," he explained, "you know they're really dumb when even the navy won't have 'em."

I paid for the beer and walked home, musing that the story would probably spread like wildfire and that Dave and Bill's Big Electrical Adventure would soon be immortalized in the lore of sleepy Millaa Millaa.

The next day Tim and I started clearing trails through one of the area's rainforest patches. Known as Millaa Millaa Falls for its spectacular hundred-foot

Our backyard in Millaa Millaa. Note the open aluminum traps—just cleaned and left to dry in the sun. PHOTO BY CHRISTINE BUEHLER.

waterfall, it was one of the largest rainforest fragments remaining on the tableland, spanning about 200 acres. We needed the trails for laying out trapping grids—arrays of cage traps and aluminum box-traps used for capturing small mammals. By counting the number of species and individuals we caught on each grid, we could directly compare the mammal communities in forest fragments to those in larger tracts of forest.

I'd worked in this same site briefly in 1984 and was shocked at how different it looked now. Millaa had received a direct hit from Cyclone Winifred five months earlier, and the damage to the forest was impressive. Big trees had been knocked down at frequent intervals along with branches, lianas, and huge tangles of climbing rattans that normally ascended the forest canopy. The rattans, covered in spines that closely resembled fishhooks, were much hated by the locals, who called them "lawyer vines" or "wait-a-whiles"—the idea obviously being that if you got tangled up in them you had to wait a while before you could get loose.

Even worse were stinging trees. Normally found only beneath gaps in the dense canopy that formed when the odd tree fell, they now sprung up everywhere. Stinging trees are an obscure relative of the North American stinging nettle but infinitely more dangerous (the latter part of their scientific name—*Dendrocnides moroides*—derives from the Latin root meaning "death").

Their heart-shaped leaves are coated with hollow, nearly microscopic spines that contain a powerful toxic alkaloid. Merely brushing against a stinging tree brings instant, burning pain that lasts for days. Perversely, for months afterward, the pain returns whenever the temperature changes—such as when you wash your hands in cold water—because your skin contracts and ruptures some of the tiny imbedded spines. We quickly developed a sixth sense about those distinctive heart-shaped leaves and gave them a wide berth whenever possible.

Tim and I made a decent two-man team. I was in the lead, slashing away with a brush-hook—a wooden pole with a scythe on the end—while Tim followed behind with a machete to prune my rough trail. Chopping trails is a great way to vent pent-up frustrations. We'd often bellow out a few kung-fu yells or a Tarzan scream just to get into the spirit of things. Tully would join in and start barking away madly. All good clean fun for your average rainforest ecologist.

Despite our efforts, it was slow going. We were hacking through a seemingly endless wall of tangled treefalls, limbs, vines, and other dead trash that had blown down in the cyclone. And we had to keep our eyes open, not just for the ubiquitous wait-a-whiles and stinging trees but also for widow-makers—broken limbs hanging from vines that could come crashing down on your head like a guillotine if you weren't paying attention. As their name implies, many a woodsman has died from these dangling limbs.

After two long sweaty days we finally cleared a set of trails at Millaa Millaa Falls. We began lugging traps into the forest, laying them out in small grids along a thousand-foot-long trail. Each trap was baited with an aromatic mixture of rolled oats and vanilla essence, which most mammals seem to find irresistible. The traps were set with a hair trigger, the door ready to slam shut at the slightest provocation.

We were using two kinds of traps: Elliot traps, made of aluminum and about five inches tall and wide and a foot long; and cage traps, built from heavy wire mesh and about a foot tall and wide and three feet long. The cage traps were awkward and bulky but past experience had shown they were vital. Indeed, if you forgot the cage traps you might as well just go home, because most of your Elliots would be flipped, tripped, or dragged off, obviously having provided some animal with a night of great fun.

In four hours we'd laid out and baited 120 traps. We returned home and

Chris Buehler baiting a trap with honey and oats—an irresistible combination for many mammals. PHOTO BY AUTHOR.

downed a couple of beers to celebrate our efforts. We retired early, pondering what we'd find the next morning.

It was a deep, reverberating growl, exuding pure, unmitigated malice.

"Jesus!" Tim said, backing up a bit. "What the hell is *that*?" He was staring at a gray-brown mammal in a cage trap, as big as a house cat. It had muscular jaws and haunches, with hind feet that sported wicked-looking claws.

"Impressive, isn't it?" I said. "*Uromys caudimaculatus*—the giant white-tailed rat." Tim was staring fearfully so I added for effect, "You sure don't want to let one of these buggers bite you. Boy-oh-boy, I got nailed by one in '84 and I bled for half an hour!" I elaborated with a few more gory details until I noticed Tim beginning to blanch.

The truth is that I was a bit nervous myself. They really *did* hurt, and I was the one who had to grapple with this monster right now while Tim jotted down the data we collected. Before coming to Australia I'd handled maybe two thousand small mammals in North America—the biggest being a two-hundred-gram kangaroo rat, which seemed pretty impressive at the time—but these thousand-gram terrors were infinitely more daunting.

I drew a heavy plastic bag over the mouth of the trap and unlatched the door. The rat stared at me for a second then shot into the bag like he'd been fired from a cannon. Tim jolted—these buggers were big *and* fast. He nervously handed me a scale which I used to weigh the rat.

Now came the fun part. Through a fairly complex series of maneuvers I got the rat by the neck and shoulders, being careful not to choke him. I now had him under control, sort of, I just had to watch out for his powerful hind feet, with which he was trying very hard to rake me.

The rat still growled like an enraged lion—the most un-rodentlike sound one can imagine—and Tim was edging further away, eyes bulging. "Okay," I said, "time for action." He nervously handed me the tagging pliers, which I slipped around the base of the rat's left ear, leaving a small numbered tag, like an earring. I then quickly examined the animal, Tim writing furiously as I called out its sex ("male"), reproductive condition ("very scrotal," meaning that his testicles were large), and external parasites ("mites on ears"). I then released the rat, which shot off like a rocket across the forest floor.

This whole thing had taken about sixty seconds, but it felt like ten times that long. Sweat was trickling down my sides. It was 7 A.M. and Tim was glaring at me like *What a godawful way to start the morning!*

We moved on to the next trap.

It was another white-tailed rat.

We trapped at Millaa Millaa Falls for five days in a row, arriving home each day as beat as if we'd just finished a twelve-hour shift in a coal mine. We'd caught a lot of animals, but only various kinds of rats. There were none of the rare rainforest marsupials we'd been hoping to see. Such species, I suspected, might be far more vulnerable than the rodents to fragmentation.

As well as a dozen or more white-tailed rats each day, we'd usually catch at least twenty fawn-footed melomys, a small but highly pugnacious rodent that squeaks shrilly when handled. A good climber, the melomys eat mainly seeds and fruits that they find in the forest's subcanopy. Rounding out the local rodents was the bush rat, an insect- and mushroom-eater that feeds mainly on the ground. Although superficially similar, bush rats are a little larger than the melomys and, to an experienced trapper, quite different in appearance: the melomys looks like a miniature teddy bear while the bush rat has a more sneaky, ratlike countenance. Learning to identify small mammals is something of an art. After a few years I was starting to feel fairly competent, though by no means a master.

In addition to trapping, on most nights I headed out after dinner for several hours of spotlighting. As in our trapping studies, the general idea was to compare mammal populations in forest fragments to those in the much-larger forest tracts that survive on mountainsides encircling the tableland. In this way, we could try to infer how each species was being affected by forest fragmentation.

Spotlighting is simple in theory: one hikes along quietly while scanning the surrounding forest for animals. My spotlight was powerful—at 200,000 candlepower, brighter than a car's headlights. To power the light I used a bulky battery designed for a Harley-Davidson motorcycle, weighing about twenty pounds.

An animal could be hiding in dense foliage, but in the glare of a spotlight its eyes glowed demonically. It was important to hold the spotlight at eye level because the reflective coating of an animal's eyes—which greatly improves its night vision—sends light back exactly to its source. By lowering the light even six inches, you might not see the animal at all.

As I gained experience, I began to perceive subtle differences in the eyeshines of different mammal species, which often helped in identifying an animal if I couldn't get a good look at it. In a spotlight, the eyes of Lumholtz's tree-kangaroos appear ruby red, while those of green ringtail possums are salmony pink. The lemuroid ringtail possum has a brilliant sunburst-yellow eyeshine, the coppery brushtail a dimmer orange, and so on. With experience one can even discriminate adults and juveniles; among lemuroid ringtails, for example, young animals have more silvery eyeshines, adults a more golden hue. Even small animals—geckos, frogs, and spiders—are revealed by a probing spotlight.

At night, the rainforest develops a whole new character. Luminescent fungi glow yellow-green while living carpets of fireflies dance above the forest floor. Tiny points of blue reveal glowworms—actually voracious fly larvae—suspended from silken threads as they await insects foolish enough to approach their deceptively appealing light.

Many creatures come alive after sunset. A descending whistle like a falling

TOP A green ringtail caught in the spotlight. During the day, these possums curl up on a branch and become virtually invisible. PHOTO BY AUTHOR.

BOTTOM, LEFT
A green ringtail possum, munching on fig leaves.
PHOTO BY RUPERT RUSSELL.

BOTTOM, RIGHT
A lemuroid ringtail in the spotlight. These possums are gregarious, living in family groups of three or four.
PHOTO BY GREG DIMIJIAN.

TOP This is how we normally see lemuroid ringtails—high in the forest canopy. PHOTO BY GREG DIMIJIAN.

BOTTOM Were it not for their eyes which glowed in the spotlight, green ringtails would be almost invisible in the rainforest. PHOTO BY GREG DIMIJIAN.

bomb reveals a sooty owl in the distance, calling for its mate. A crash of high branches indicates a family of lemuroid ringtails, leaping gamely from tree to tree. From the forest canopy, angry squawks suddenly erupt, heralding the arrival of a group of flying foxes—huge bats with three-foot wingspans that feed greedily on fruits and nectar. And underlying all of these distinctive sounds is the constant trilling and chirping and droning of myriad insects.

The rainforest seems almost mystical at night. Not only is the forest pulsating with life, but one's senses achieve a clarity and focus unattainable in the light of day. Spotlighters can become mesmerized by the forest, so attuned are they to the nuances of life around them. And outside the forest, the southern sky has its own special beauty. Far from any city, the brilliant lights of the Milky Way and Southern Cross are undiminished by earthly competitors.

Tim preferred to stay home at night, so I often took Tully out spotlighting, both for the company and because he loved the walks in the rainforest. He was turning into a first-rate field companion, following on my heels and keeping wonderfully quiet. This was in stark contrast to his general behavior around the house, which he regarded as his own personal playpen. Tully was learning to use his big brown eyes strategically, adopting a heart-melting sad-puppy look whenever Tim or I tried to show him the door. Even the neighbor kids in Millaa were falling for his act, and would knock on our door and politely ask, "Can Tully come out and play?"

One night I was spotlighting with Tully at one of our fragments, a rainforest patch of about thirty acres surrounded by rolling cattle pastures. We'd heard a few dingoes howling eerily in the distance, and as a result Tully was sticking very close to me. There were no roads bisecting the patch, so we were conducting our census by hiking around the patch's perimeter, slogging though the rough grass of the cattle pasture. My mind was focused on the forest, consumed with the task of finding animals. I'd walk parallel to the forest edge for a few paces then stop and turn, sweeping my light back and forth across the solid wall of forest in front of me. I'd seen very few animals that night, though my senses were attuned to even the tiniest flicker of an eyeshine.

Suddenly Tully tried to crawl right between my legs. I stumbled a bit then turned around—and my heart nearly stopped. Dozens of huge yellow eyes were staring at me, only yards away. I let out a startled yell and Tully started barking frantically, bravura and fear all coming out at once.

The cows turned and bolted off into the night.

"Dammit Tully!" I yelled, "You scared me half to death." Then I started to laugh. Somehow a whole herd of dairy cattle had snuck up behind us, their curiosity piqued by the strange nocturnal goings-on in their paddock. For the remainder of that night we had to deal with those crazy cows, who tried to sneak up on us again and again. I'm sure it was the most fun they'd had in years.

friendly leeches

It was our third week of fieldwork and two things happened almost simultaneously which would later have a profound impact not only on me and my project, but also on the normally placid social dynamics of Millaa Millaa.

First, a Swiss traveler whom Tim and I had met in Cairns, Christine Buehler, unexpectedly showed up in Millaa. She had been intrigued by our project and was now keen to help us if possible. We quickly accepted her offer.

Second, on the following day Tim, Christine, and I were driving down a country road and saw a lone backpacker, forefinger extended to hitch a ride. We picked him up and learned that he was from Melbourne. We began telling him about our research project and in the space of only a few miles Richard, too, abruptly volunteered for our little team.

I was ecstatic—in two days I'd doubled my field crew—and we needed all the help we could get. Tim and I had been working frenetically six days a week, but progress was painfully slow. Everything seemed to take twice as long as I'd planned in the cozy confines of my student office in Berkeley. My goals for the project were ambitious, and until now we'd had little hope of achieving them.

Richard and Christine couldn't have had more different personalities. Christine was fresh-faced and bubbly, a constant source of melodic laughter and an incredibly hard worker. It was impossible not to warm to her immediately. Richard, however, was a craggy, sarcastic character, a hard-drinking Irish product of working-class Melbourne. Richard fit the Australian definition of a "good bloke" and I liked him, but his interests ran more toward having a wild ol' time than a pristine wilderness experience.

There were a few other complications with Richard. For one thing, he and Tim began butting heads almost immediately. Tim never really grasped the notion that if an Aussie is cruelly teasing you, that generally means he likes you. By becoming defensive, Tim was violating an unwritten code of Australian conduct. Richard soon decided that Tim was a "wanker"—literally a masturbator but actually connoting a fool—and proceeded to lace him with an endless barrage of sarcasm. Tim took all this rather gamely, I thought, and did his best to reply in kind.

In addition, Richard was an avid pot-smoker. In Millaa Millaa it was perfectly acceptable to get falling-down drunk six nights a week, but smoking marijuana bore obvious risks. The social tension between the few odd hippies that lived in these parts and the rest of the populace was palpable. In 1984, the ultra-right-wing premier of Queensland, Sir Joh Bjelke-Petersen, had ordered a forty-mile road bulldozed through the last major stretch of lowland rainforest in north Queensland. One of his chief justifications was that there had been a nest of pot-growing hippies living up in the remote wilderness and the police needed to get in there and flush 'em out.

While I personally held more liberal views than the reactionary premier, it was clear that pot-smoking in Millaa Millaa wasn't worth the risk. Moreover, we had important work to do, and I surely wasn't going to risk compromising all our efforts just so Richard could get stoned. So I laid down the law. Richard reluctantly agreed to forego marijuana and to hide his stash someplace off my property—I didn't want to know where.

Despite my efforts to keep our collective noses clean, the Millaa Millaa policeman soon became convinced that I was actually a major drug kingpin leading what was obviously an international conspiracy to flood the streets and innocent youth of his little town with illicit drugs. No doubt my friendship with Geoff Downey, that infamous pot-smoker, had helped to reinforce this notion.

Our first confrontation with the local cop occurred only days after Richard and Christine joined us. We'd been clearing trails in an area called Lager Bend (so named because a beer truck had once careened off the road there), a large tract of rainforest ten miles south of Millaa. We'd spent an exhausting day deep in the forest and were just returning to the road, muddy and laden with machetes and brush-hooks, when the Millaa cop suddenly popped out from behind a tree.

We were speechless.

"How ya goin', mate?" the cop inquired to me, his severe tone and expression revealing that he considered me anything but his mate.

"Uh, well, uh, we're okay, I guess. Um, how are you?" I stammered, completely off balance.

His manner became even more suspicious. Obviously I was nervous about something. "Those your star-picket posts up the track?" he fired quickly, jabbing a thumb behind him.

I had absolutely no idea what he was talking about. "A star what . . . ?"

He stared at me hard, like a celebrated detective sizing up a mass-murder suspect. Then he repeated slowly "*Star-picket post*, mate. I asked if those were *your* posts up by the road."

This was actually getting scary. I still hadn't the faintest clue what he was talking about and Richard, the lone Aussie among us, wasn't helping a bit. The cop obviously thought we were returning from a long day of alternative agriculture in the rainforest, and the fact that he figured I was clumsily playing dumb wasn't helping matters.

He looked us up and down with our muddy clothes and lethal machetes and brush-hooks and seemed to be mulling over whether to yank out his gun and run us all in on suspicion of drug production. "You just watch your step around here, mate," he finally said to me. "I've got my eye on you."

With that, he pivoted and stalked off toward the road, ducking and weaving among the vines and buttresses in his black street shoes and tidy blue uniform.

The four of us stood there for a long moment, recovering from the shock. Finally I said to Richard, "What the hell was he talking about—all that stuff about star-picket posts?"

Regaining his voice now that the cop was gone, Richard explained that a star-picket post was a metal fencepost, shaped like a plus sign—a star—when viewed from above. I remembered seeing one that morning near the logging track, probably left there long ago by some logger.

We started trudging toward the road. Suddenly Richard turned to me with great enthusiasm. "You know, Bill, you really *ought* to think about growing a crop out here. You've got the perfect excuse to be in the forest and that clueless ass"—he nodded at the departing cop—"would never catch us."

I gave Richard my most withering look until I realized the bastard was pulling my leg. He cackled joyfully as we hiked back to the truck.

❧

Including our mascot, Tully, we now had a team of five, and our efficiency increased proportionally. We were developing a good rhythm in the field, working like a trained surgical team as we checked our traps each morning. I handled the animals, Tim handed me eartags and other gear, Richard re-

A giant white-tailed rat, still groggy from halothane. The gas anaesthetized the animals for a minute or two and made trapping a lot less stressful, both for them and us. PHOTO BY AUTHOR.

set the traps, and Chris recorded the data—the name of each species, its sex, weight, and tag number, and other details. Tully had been banished from trapping after tearing off after a giant white-tailed rat, though I still wasn't convinced he was actually planning to tangle with it.

We'd begun using halothane gas to subdue our animals. A bit like nitrous oxide—laughing gas—halothane made things a lot less stressful, both for the animals and for us. Rather than endure a series of knock-down-drag-outs with the white-tailed rats every morning, we simply shoved the animal into a plastic bag full of the gas, waited about fifteen seconds, then extracted and processed it as usual. It was easy. We also could get a much better look at the animals, some of which—like the bush rat and Cape York rat—are so similar they often confuse even the experts.

Halothane numbs an animal for about sixty seconds, which was perfect for our purposes. I'd release the animal just as it was waking up, and it would stagger like a drunken sailor back to its den. The fact that we commonly caught the same animals over and over showed that the halothane didn't bother them. Indeed, when we recaptured an animal Richard often quipped, "It's bush rat number 1202, back for his daily hit of halothane!"

Christine was tireless. While the rest of us returned from the forest each day only to collapse in an exhausted heap—our sole priorities being food, shower, and a nap, in that order—she would build a roaring fire to dry our clothes then run around the kitchen making hot chocolate and toasted-cheese sandwiches. Her only real deficiency was cooking; she seemingly could burn water, and her carbonized sandwiches were fast becoming legendary. She was cute—slender with chestnut-brown hair—and Richard obviously adored her, though it would have been un-Australian for him to let on about it too much. Instead, he ribbed her constantly, a treatment she endured with a cheery laugh that seemed to light up the room.

Our ritual upon returning home from the field each day including stripping off our clothes to search for leeches. Terrestrial leeches abound in the Australian rainforest, especially on rainy days. With its hyperwet climate, Millaa Millaa is leech heaven.

The brown rainforest leeches are phenomenally aggressive. Attracted to movement, body heat, and carbon dioxide from your breath, it is virtually impossible to deter them, even with liberal doses of insect repellent. The mouth of a leech contains three razor-sharp jaws, which it rotates back and forth to create a perfectly circular hole in the skin surface. To improve the flow of blood, leeches inject an anticoagulant, making leech bites bleed—and itch—profusely.

One of the more famous leech legends concerns Christie Palmerston, the first white man to explore the Atherton Tableland. After a hard day's hike, Palmerston lay down under a tree for a nap. When he awoke his nose was completely plugged up—with bloody, engorged leeches. The locals love relating this story to tourists and prospective biologists.

We were quickly accumulating some leech stories of our own. In 1984, I'd gotten a leech on my lower lip while trapping, and decided it would make a great seminar photo to gross out my fellow graduate students back home. Unfortunately, my camera was in my truck, and I still had a half hour of trapping to go before I could leave the forest. I endured that damned leech for twenty minutes, feeling it growing fatter and more engorged by the minute. Finally I could stand it no longer—I yanked it off and threw it into the forest, my lip bleeding like I'd been stabbed with an icepick.

My other horror story—one that I haven't shared with many people—also happened in 1984 when I'd been camping in wet sclerophyll forest. The forest was swarming with tiger leeches—ugly greenish beasts with a yellow stripe down their backs—which became as big as a man's thumb when engorged with blood. Going to the loo under these circumstances was the ultimate challenge; as one squatted in the bush, one could almost hear dozens of leeches approaching, racing across the forest floor like eager inchworms. Defecating in the forest became a sort of perverse race between the leeches and my bowels, which were not performing as well as usual, given the highly pressurized circumstances. One day a tiger leech managed to secret itself inside my pants. Later, as I hiked through the forest I noticed a distinctive itching in my private parts. I immediately dropped my pants and was aghast to find a massive leech attached to my willy, half its body inside the urethra. I howled in horror as I yanked the engorged leech out, then endured days of intense itching afterward. The only highlight from this appalling incident was that for the following week my private parts swelled up hugely, which was gratifying from a purely egotistical point of view.

❧

While mammal trapping was intense and grueling work, the spotlighting proved more enjoyable. The only hard part was forcing myself to stagger out into the night after a big dinner and sometimes a beer or two, leaving our cozy house behind. Christine was keen to try spotlighting and on her first night found herself almost face to face with a big Lumholtz's tree-kangaroo. She was entranced. After that she wanted to go spotlighting every night.

Tree-kangaroos are odd beasts. Closely related to wallabies, they are a testament to the Rube Goldberg ingenuity that Mother Nature sometimes employs when the proper materials just aren't at hand. In a continent without monkeys, the arboreal niches in Australia were relatively open. So the ancestors of tree-kangaroos—which like the wallabies were highly evolved terrestrial hopping machines—decided to climb trees.

The tree-kangaroo differs in many ways from conventional kangaroos and wallabies. Its tail is much longer than a kangaroo's and is used as a counterbalance while climbing. Whereas normal kangaroos have stiff, narrow hind feet, the hind feet of tree-kangaroos are broad and flexible, almost like a human's. Tree-kangaroos also have short ears and muzzles, giving them a monkeylike countenance. Finally, they have long, muscular forelegs for gripping trees, with heavy claws on each paw.

Though they spend much time in trees, tree-kangaroos are far from the most elegant of climbing animals. Their ambivalence toward climbing is perhaps best illustrated by their escape response. When threatened, they do the exact opposite of most arboreal animals—rather than climb higher, they plummet to the ground and bound off at high speed. These leaps are sometimes spectacular. I once saw a startled tree-kangaroo plunge fifty feet to the ground, landing with a thunderous crash that was followed by silence. Surely, I thought, it must be badly injured, but when I approached, it bolted into the forest in the usual manner.

One night we went spotlighting on Mt. Fisher, a large forest tract southwest of Millaa Millaa. Tim had come along with Chris and me, foregoing

The Lumholtz's tree-kangaroo, an adequate but not very elegant climber. PHOTOGRAPH BY RUPERT RUSSELL.

his usual card games and bantering with Richard. To reach the forest we'd driven south until we came to a climbing dirt track which we followed for a few miles, passing an old shack just before the forest. A light burned inside; obviously it was inhabited.

We had a pleasant night at Mt. Fisher, recording five species of mammals, a good total for any night. We climbed wearily into our truck and turned for home, and had reached the paved road when a car suddenly came blazing down the track behind us.

As I turned onto the road the car pulled right up behind us and stayed there, its lights flashing onto high beam. This was aggressive and unusual behavior; traffic was sparse in these parts and people usually gave each other a wide berth. The car followed right on our tail all the way to Millaa Millaa. I was tempted to pull over and see what this guy's problem was but we were feeling pretty nervous. It was well past midnight. Was he crazy? Was he armed? Tim was clutching our only weapon—a machete.

At Millaa we'd decided we'd had enough. I lurched to a stop in the middle of the main street and turned off the engine. We sat there, tense as coiled springs. The car stopped and the driver jumped out, his high beams still brightly illuminating us.

He stalked up to the truck, obviously angry and intensely suspicious. He stuck his head in my window and looked around inside as if searching for something. "What the hell were you doing up there in the rainforest?" he fired at me.

"Just spotlighting, mate," I replied, much more calmly than I felt. "We're biologists, studying the mammals in this area." I described our project but the situation was extremely tense—we were being interrogated. Equally puzzling was that this guy was a hippie, bearded with long dark hair in a ponytail. In Millaa Millaa a farmer might be induced to run you off his land, but the hippies were rarely so territorial.

The conversation wasn't going anywhere: one o'clock in the morning and we were having a bizarre standoff on the main street of Millaa Millaa. Finally I suggested he follow us home, where I could prove I was telling the truth. There, over a coffee, the tension slowly drained out of him as we elaborately described our project and what we were trying to do here.

His name was Franco. He was about thirty years old, olive-skinned and handsome. He explained that he thought we were pot-growers. He was worried that if someone found our crop growing near his shack they'd assume it was his. Having experienced the paranoia of the Millaa cop at first hand, I found this explanation eminently believable. Franco finally left, obviously mollified. Chris, Tim, and I all sighed and looked at each other, our tired expressions conveying the same thought—*What a loony night!*

I met Franco on several other occasions and got to know him reasonably well. He was known as a hot-tempered Italian, a description I couldn't have disputed. He'd come to north Queensland several years before from down south, traveling up the coast with his beautiful wife on a sailboat. They now lived with their baby daughter out in Franco's shack abutting the rainforest. We passed by the shack many times on our way to Mt. Fisher, but after the drama of our first meeting I always beeped my horn, just to let Franco know it was us.

Some funny rumors swirled around about Franco, which led me to believe he hadn't been forthcoming with us on the strange night we first met. One rumor suggested he was a major pot-grower himself, that he put bread on the table by shipping his crop off to the big cities down south. Geoff Downey explained with a wink that that was why Franco was so territorial about his rainforest—he'd thought we were trying to steal his crop. Apparently that kind of thing went on more often than people realized, and violent clashes sometimes ensued.

Later events would convince me that Geoff was probably right. One night Franco simply disappeared. Missing for a day, his worried wife hiked up into the rainforest and eventually found him, dead, his belly ripped open by a shotgun blast. The police determined that he'd been shot a half mile further up the track; he'd been desperately trying to crawl home when he died. His distraught wife took their baby and left for some destination down south. The police never caught Franco's killer.

Before she departed, Franco's wife left him a final memorial. A big cement tank had been constructed by a local farmer high up on Mt. Fisher—deep in the rainforest near where Franco had died—to provide water pressure for his farm below. There, Franco's wife had painted a simple epitaph: Franco's name, birth date, and death date in giant black letters. I often passed that tank at night, and the back of my neck sometimes tingled as I wondered whether Franco's spirit might be watching me, still jealously guarding his rainforest.

July was our second month of fieldwork, and we were entering the driest part of the year—though in a soggy place like Millaa Millaa the term "dry season" is relative. Despite Tim and Richard's squabbling, things were pretty harmonious.

We'd been trapping in a large forest tract at a place called Mt. Father Clancy. Near the mountain's summit, the effects of the recent cyclone had been devastating. Huge trees had fallen and with them masses of lianas and wait-a-whiles. Stinging trees grew everywhere, and as we slashed our trails we sometimes found dense clumps of stingers right where our trapping grid was supposed to go.

Not only are stinging trees dangerous to touch, but when we cut down those directly in our path we often inhaled the minuscule spines. The spines caused an intense burning sensation, making our eyes water and noses run profusely. Sometimes we'd have great sneezing fits. Once the pain was so bad I couldn't drive home; I had to pull over and lie down for an hour.

At times Richard thought I was completely mad. "Why the hell don't we just put the trapping grid over *there*?" he'd demand indignantly, pointing to a nearby site that was completely clear of vines and stingers. When I tried to explain that this would be unscientific—putting our grids wherever we felt like it would be a biased and subjective way of sampling the forest—he'd reply with great seriousness, "Look, Bill, we won't tell *anybody* about this, will we?"—a statement followed by sharp looks to Christine and Tim to ensure their compliance.

Of course I refused to budge on this issue. Richard would then complain vociferously about what a bunch of wankers biologists were.

Our trapping at Mt. Fisher was quite exciting. We were just entering the breeding season of antechinuses, and it was here that we encountered the first of these strange beasts.

I knew we'd caught something new when an irritated growl—a kind I'd never heard before—emanated from a trap. The trap was surprisingly light, the sound wholly disproportionate in volume to the animal's diminutive size.

I extracted the animal and it promptly bit me, latching on to my finger and hanging on with the ferocity of a badger. It was amazing: it simply refused to let go, so enraged was it by the indignity of being captured. And for such a small animal—only a bit bigger than a field mouse—it packed a pretty impressive bite.

As Christine tried to extricate my finger by prying the animal's jaws open, I thought about the antechinuses, legendary among biologists. In the same family as Tasmanian devils—a marsupial version of the wolverine—antechinuses are renowned for their fierceness. These tiny marsupials have dozens of sharp teeth and think nothing of tackling and killing an animal twice their size. Ounce for ounce, perhaps only the shrew is equally voracious.

But despite their legendary aggressiveness, antechinuses have another feature that is even more intriguing—one of the oddest life histories of any mammal. Male antechinuses live less than a year, and toward the end of their short lives they become phenomenally stressed, so much so that they neglect to eat or sleep. Like migrating salmon, they become utterly consumed by their mating drive. They copulate incessantly, attacking any other males in sight, and then, within a week or two, every male dies, their immune systems overwhelmed by the stresses of the preceding weeks. For nearly a month each year, there is not a single male antechinus in the world—only pregnant females.

As we examined the antechinus I described their bizarre mating and dying frenzy to my companions. Richard grew quiet for a moment then quipped, "Those poor bastards wouldn't know if they're coming or going!" Not a bad pun, though we all groaned loudly.

Having captured our first antechinus, we decided to take it home for a few days to observe its behavior. Tim and Richard went to work building a big

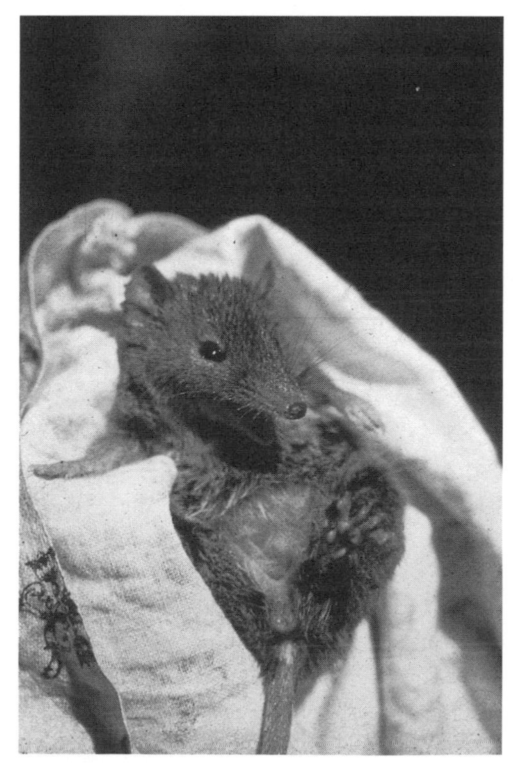

TOP Yellow-footed antechinuses become more abundant in forest fragments, apparently because they prefer disturbed habitat. PHOTO BY MIKE TRENERRY.

BOTTOM A brown antechinus. Like badgers, these tiny terrors would sometimes latch onto our fingers and refuse to let go. PHOTO BY AUTHOR.

terrarium, using some plates of window glass we'd found in one of the sheds.

They were trying to cut the windows to fit their design but I kept hearing the sounds of breaking glass, followed by Richard bellowing at Tim. Things obviously weren't going well. I thought about trying to intercede but figured, what the hell, they're big boys. Finally Richard came storming into the house, his face flushed with anger. "That goddamned Tim couldn't organize a root in a whorehouse!" he roared ("root" being Aussie slang for sexual intercourse). I couldn't help but laugh. Poor Tim. God help anybody Richard decided to zero in on.

We finally got the terrarium built and the antechinus settled in nicely. He immediately came out of hiding to roam among the rocks and grass of his enclosure. He absolutely loved moths, and soon learned that the sound of the lid sliding back meant dinner was served. He'd wildly chase the moth until he caught it, then sit back on his haunches and munch away happily, holding it in his paws like a hamburger.

The weather had been sunny and dry on Mt. Father Clancy, and expecting more of the same, we were dressed only in T-shirts and shorts. But around 9 A.M. the sky started to darken. The wind picked up quickly and began whipping through the treetops. Fat droplets fell, and within moments the deluge hit. The temperature plummeted as dense sheets of rain poured down upon us, and we were quickly soaked to the bone.

We were only half finished with trapping. Chris was desperately trying to keep the data notebook dry by shielding it with her body. The storm showed no sign of letting up.

Richard kept looking at me anxiously and finally yelled above the storm, "We've got to get the hell out of here!"

I shook my head. "We can't—we've still got animals in the traps."

"Let's just let 'em go! This is nuts!"

I shook my head again. No way. Releasing all the animals would screw up a whole week of trapping data—an incomplete census would be worthless for comparing this site with others. We'd just have to go as fast as we could. I wasn't really worried about the animals at this point because the traps provided shelter from the rain.

"This is crazy, goddamit!" Richard yelled again. "Look, Bill—Chris and Tim want to go home too."

I looked over at them and couldn't deny it—they were like the proverbial drowned rats. But despite Richard's efforts to incite mutiny, we couldn't simply give up and go home every time the weather turned bad.

Then Chris suddenly looked inspired. She spoke to Tim for a moment, who turned and jogged off toward the truck. In minutes he was back, a big tarp in hand. As the rain thundered relentlessly down, we wrestled the tarp into an elongated rectangle which we draped over our heads. Like a bizarre giant caterpillar, we slowly marched off in unison to finish checking the traps.

Later, I thought about how tempted I'd been to give in to Richard and just let all the animals go. I was glad we hadn't—we'd faced the worst of the local elements and hadn't given up. To Richard, however, this incident simply reinforced a growing belief that field biologists were dangerous lunatics. He informed me that I was the stubbornest bastard he'd ever met, hiding just a hint of a smile as he said it.

the mystery of the day-glo rats

"Bill, there's something I have to tell you. Richard is sniffing halothane."

I groaned—bloody Richard! I looked at Tim. "Where is he?"

He pointed to the woodshed. We walked over and sure enough, Richard was inside with a guilty grin on his face. His face was flushed, but not from embarrassment. He was holding a plastic bag with a few ounces of liquid halothane in it. He'd been sticking his nose into the bag and inhaling the fumes. I just stared at him, utterly at a loss. What was I going to do with this guy?

"Hey," he said defensively, "if it's good enough for the rats it's good enough for me. You said no pot but you didn't say anything about halothane."

I wasn't sure whether to laugh or throw a screaming tantrum. Heck, I'd once taken a whiff of halothane myself, out of curiosity. It produced a pleasant feeling of lightheadedness. But sniffing the stuff like it was glue or something? Unbelievable.

I held out my hand. Richard stared at me for a long moment, then shrugged and handed over the bag. "Weird, Richard. Altogether too weird." I shook my head as I walked back to the house. Richard had been here a month and recently had said he needed to be moving on. I liked the crazy bastard but had just decided I wouldn't try to talk him out of it.

Despite my annoyance, as Richard's departure approached I started feeling blue. I'd gotten used to his crude jokes and irreverent view of the world. Tim would be leaving also, flying back to California in two days, so the four of us decided to drive down to the coast and camp overnight on a beach near Cairns. I arranged for the neighbor kids to feed Tully, then we crammed into the truck and took off.

We bought wine and beer and had a debaucherous night on Holloway's Beach, north of Cairns city. A rather interesting dynamic developed that evening. Richard was obviously holding a torch for Christine, but Christine, to my surprise, actually seemed keen on me. Against my better judgment I ended up smooching with her that night. Richard pouted briefly but soon became diverted by a wild political argument with Tim.

We awoke at dawn feeling universally awful. I'd always suffered from killer hangovers, and gritty sand seemed to have penetrated my ears, mouth, and most other orifices. Only slightly rejuvenated by a swim in the sea, we packed up our things and drove into Cairns.

We all traded hugs as we left Richard and Tim at the youth hostel. Christine was returning with me to Millaa Millaa. She and I were unusually quiet as we climbed the winding road back up to the Atherton Tableland. I kept thinking how the four of us had formed a kind of surrogate family—a loony family, of course, but then many families are. Though Richard and Tim had been with us only for a month, it felt like we were returning to an empty nest.

We were back to a team of two. Chris was a terrific worker but our efficiency had still been considerably reduced. At home, things were quiet—almost domestic. The morning after returning from Cairns, Chris woke me with a cup of tea and we'd ended up embracing. We nearly took the next step, but something made us hesitate. For my part, I hadn't been able to stop thinking about Beth, and wasn't sure I was ready for an intense new relationship. I also liked and respected Chris too much to even contemplate a casual affair, and knew she couldn't stay in Australia much longer because her visa would soon expire. By tacit consent, we decided to keep our relationship platonic, at least for now.

Chris had few limitations, but one thing she couldn't do was drive a car. In Switzerland, trains and buses go almost everywhere, and many people don't ever learn to drive. Normally this wasn't a problem, save for one complicated day.

We'd been working in a remote area behind Mt. Father Clancy. To reach our study site we drove twenty miles south of Millaa, then gingerly crossed a decomposing wooden bridge to reach an old logging track. We followed the track for another five miles, then parked and hiked for twenty minutes up a rocky mountainside to reach our traps.

This sort of thing was pretty much par for the course, but this week we'd been inundated with rain, and the track had become a slippery quagmire. Our tires were fully caked in a layer of mud, obscuring the tread and leaving us with little traction. Several times I'd nearly lost control and come dangerously close to sliding off the track, which fell off sharply below.

After a long day of trapping we'd returned to the truck tired and muddy. I began a series of tight maneuvers to turn the truck around on the narrow track. Suddenly we began veering sideways toward the edge and a fifty-foot vertical drop. I slammed on the brakes and suppressed a yell as we skidded to a stop only inches from the ravine. Our front wheels had actually fallen over the embankment and we were now up to our axles in mud.

Our relief at still being in one piece was short-lived. We were tired, bogged, and a very long way from anywhere. With a sigh I grabbed a shovel from the back and began digging at the mud while Chris cut branches to stuff under the tires. After ten minutes of effort I climbed behind the wheel and tried to reverse back from the edge, while Chris pushed from the side with all her might.

We hardly budged. If we'd had Richard and Tim we might've gotten out but Chris just wasn't strong enough, and I had to drive the truck. The afternoon was slipping away and I was feeling desperate. I told Chris she'd have to drive so I could push; maybe it'd be enough. She looked at me as though I'd lost my senses—she *couldn't* drive, she didn't know how. But I was adamant; we had few options.

She grew very quiet as she climbed behind the wheel. I showed her what to do, releasing the clutch and revving the engine so the rear wheels would turn but not spin wildly. After killing the engine a few times she began to get the hang of it while I heaved and swore as my feet slipped in the mud.

But our efforts were to no avail. The truck hadn't shifted more than a few inches. I was getting madder by the minute until I saw Chris quietly sobbing. I was taken aback, it was so out of character. I later learned that she was petrified of driving. Her compulsory driving lesson here on a dangerous precipice had completely unnerved her.

Grudgingly conceding defeat, we began our descent down the mountain. The road was a long way off, and nightfall was rapidly approaching. As we walked, we said little.

We reached the road well after dark. It was a relief to get off the boggy track and onto solid pavement, almost like stepping off a swaying boat onto solid land. In the distance we could see a faint light, and set off in that direction. We finally staggered up to a farmhouse like a couple of bedraggled, muddy hobos.

On closer inspection we could see it was less a farmhouse than a faded wooden shack with an aluminum roof. Littering the yard was an odd collection of rusting farm machinery, old cars perched on wooden blocks, a discarded washing machine, and various other accoutrements.

We knocked on the door and it was quickly yanked open by a short, swarthy man with wild-looking eyes and scraggly beard. He looked us up and down then bellowed, "What'll you be wanting, then?"

We quickly explained who we were and why we looked so awful. We recounted how we'd gotten bogged, and said we needed help to excavate our truck—did he by any chance have a tractor? He digested all this without comment then said, "Oh, all right. But I'm in the middle of tea right now so you might as well join me."

We shed our muddy boots and followed him inside. The interior was as festooned with accumulated junk as the outside: ancient magazines, phonograph records, hand tools, old knickknacks. We sat down at a dinner table while he disappeared into the kitchen, reappearing with a six-pack of beer. My eyes brightened. He tossed me a beer and indicated for us to help ourselves to food. Then he sat down just inches from Chris. She stiffened a bit, but didn't edge away. We attempted conversation but were mostly focused on eating and resting our aching feet.

He seemed happy to have some company. He explained in a rough Irish brogue that he lived alone on his farm. His great passion was his motorcycle and he grew animated as he described how he could make the twisting run to the coast in only an hour—an impressive feat, if true. He told several increasingly bawdy jokes and laughed uproariously, edging even closer to Chris. She looked uncomfortable but stoically kept her composure, staring down at her plate.

Our host turned away abruptly and did something to his face. As he turned back he held his hand right next to Christine's ear, then tapped her on the shoulder. She turned to look and shrieked. I jumped up as he collapsed in laughter, pounding the table and waving his fist around. He opened his hand and there was his eye. The lunatic had a glass eye, as lifelike as you could imagine.

Christine gradually recovered her composure but clearly had lost her appetite. Thankfully, dinner drew to a close without further hysterics. We climbed onto our host's tractor and in twenty minutes had reached our truck and dragged it free from the mud. We followed him down the mountain and I waved and beeped my horn in thanks.

As we headed toward Millaa I asked Chris in a deadpan voice if she wanted to go back to his house now for dessert. She replied with a very rude gesture that she must have learned from Richard.

We hauled our two-hundred-odd traps back to Millaa Millaa the next day, then began the least pleasant part of our ritual: washing the traps. The cage traps just had to be hosed down and dried in the sun, but the Elliots needed to be carefully cleaned; otherwise the bait and mammal droppings inside spawned an impenetrable mass of fungi.

We disassembled the Elliots by extracting a long wire that linked the floor and one wall; the trap then fell open. The traps were soaked with a garden hose, scrubbed with stiff brushes, then hosed down again. We never used soap, just water—a lesson I'd learned the hard way as an undergraduate. Our mammalogy class had set up a big trapping grid in the Idaho desert, but we didn't catch a thing. The mystery was eventually solved when we learned that a lab assistant had washed the traps in dish soap. Although thoroughly rinsed, a faint soapy smell was still present on the traps—undetectable to us but apparently reeking to the mammals. It was an important lesson in mammal olfaction.

As the traps dried in the sun, we began fixing several damaged ones. Captured mammals sometimes take their frustrations out on the trap itself, gnawing away at the treadle or doors. The power of their jaws is impressive, and chewing through aluminum seems relatively easy for them. The whole operation—cleaning, drying, repairing, and reassembling all the traps—took us eight hours.

The following week bore witness to one of my oddest experiences as a mammal trapper. So much so that I became convinced we were the victims of an elaborate practical joke.

We were at Millaa Millaa Falls, our 200-acre forest fragment, and were mystified to discover one morning that virtually all of our Elliot traps had been opened. The traps were not only open, they were shiny clean, not a trace of bait inside. Opening an Elliot trap isn't easy—Chris and I used pliers to tug them open—and we couldn't fathom an animal doing that. In each case the long wire that held the trap together had been removed and was lying nearby on the ground.

This was beyond my realm of experience. For a moment the theme of *The Twilight Zone* played in the back of my mind. Surely one of the local yobbos must be pulling our legs. Aussies love practical jokes, generally the meaner the better. But I just couldn't imagine someone going to the trouble of scrubbing dozens of traps clean, removing every trace of bait.

Chris and I reassembled and rebaited all the traps, then went home. That afternoon we drove back to the fragment to see if any cars might be parked nearby. We saw nothing.

The next morning it happened again. Virtually all the Elliots had been opened. I was slowly coming around to the idea that an animal must be doing this. To start with, I couldn't imagine someone going to so much trouble for a joke. In addition, the traps had apparently been licked clean, and I was quite sure no sane person would do *that*.

This was clearly a test of our wits and determination—we *had* to solve the mystery. We assembled and baited the Elliots yet again, then went home for lunch. When we returned we were armed with additional cage traps. Unlike the Elliots, the cage traps had been functioning normally. We laid out the extra traps and headed home again, eager to return the next morning.

As expected, nearly all of the Elliots were wide open, but many cage traps held various kinds of rats. Now we had a trick up our sleeve. Christine was carrying three plastic bags, each containing a bright fluorescent powder—red, blue, or green in color. The powder was nontoxic and superfine, like talcum. When illuminated by a handheld ultraviolet light, it glowed brilliantly.

I'd originally intended to use the powder to study the foraging habits of mammals: they'd leave behind traces of dye as they fed, and we could infer what they'd been eating. But today it would be put to a higher purpose. After tagging each animal, we dropped it into one of the bags and gently shook it. The animals emerged in the most stunning Day-Glo colors—the term "punk rats" immediately sprang to mind. We dyed all the melomys green, all the bush rats blue, and all the white-tailed rats red.

Now we had a fighting chance. Though it was costing us time, this was fun. I imagined myself as Sherlock Holmes, pondering some complex Victorian murder case. We'd call this one *The Mystery of the Day-Glo Rat*.

We returned the next morning and sure enough, the conundrum was solved. The Elliots had again been opened, and nearly all bore traces of red dye—the white-tailed rats were our culprits. We weren't sure if a single animal had been causing all the mayhem or if he was teaching his friends and neighbors how to open traps as well. The white-tailed rats had earned our grudging respect. They'd learned—probably by trial and error—to solve a problem that would have taken some people I know a fairly long time to figure out. Obviously they were far more intelligent than we'd realized—a worthy adversary for any mammalogist.

As we climbed into the truck Chris looked at me. We really should be prepared, she said, for a phone call from some hysterical local resident, screaming that there was a fluorescent blue rat in her pantry.

At the end of our strange week Chris and I took a day off for birdwatching. The rainforest is a difficult place to see birds, but every morning we heard a cacophony of calls—the sharp whip-cracks of eastern whipbirds; chowchillas sounding like unearthly electronic pinball machines; the haunting, babylike cries of spotted catbirds.

It was a beautiful day, and we decided to tour some of the tableland's waterfalls and birdwatch along the way. Kookaburras laughed like raucous monkeys in the treetops. Flocks of rainbow bee-eaters—brilliantly colored acrobats—chased bees and dragonflies. In a flowering tree, crimson rosellas chattered loudly as they drank fermented nectar. The birds were tipsy—several had actually descended to the ground and were almost falling-down drunk. The rest were carrying on exuberantly, clambering wobbily along branches to reach more flowers.

The phone was ringing as we walked in the door. It was Richard, calling from Cairns. He'd had an inspiration: why not advertise for volunteers in the local youth hostels? I obviously needed all the help I could get and it'd be the adventure of a lifetime for the travelers.

I mulled this over. Why not?—maybe we'd get a few volunteers. I thanked Richard for his idea and jotted down his address at the hostel. The next day Chris and I made up some colorful flyers, describing our project and the many opportunities to see exotic wildlife. We highlighted the beauty of the Atherton Tableland and the mystique of the rainforest, but pointedly neglected to mention stinging trees, leeches, and wait-a-while vines. We mailed off the fliers to Richard, who'd promised to distribute them to all the hostels in Cairns.

In retrospect, this was unquestionably Richard's finest moment. At the time I failed to appreciate the significance of this small event, but it would soon transform our project completely.

invasion of millaa millaa

I'd gone off to visit Geoff Downey for the afternoon and returned to find an exasperated Christine. The phone had begun ringing at noon and hadn't stopped since. American travelers, English travelers, German travelers—they all desperately wanted to come to Millaa Millaa to work on the project. She'd been telling them to ring back tonight and speak with me. As she was explaining all this the phone rang; it was a South African woman who thought the project sounded absolutely perfect—could she and her boyfriend come up immediately?

In this way our project was transformed almost overnight from a fledgling two-person operation to a rollicking ten-person mob. In many ways the timing was good. I'd been in Millaa Millaa long enough to get settled in, and Christine's visa had nearly expired and she'd soon have to leave Australia.

Our new crew consisted of a French Canadian, Patrick; an American, Rich; a German, Joachim; two British women, Gail and Sue; a South African, Cathy, and her American boyfriend, Dale. In addition to these seven, John Vollmar, a recent graduate of UC Berkeley, had just arrived from California. I was paying John's travel and living expenses and in return he'd be working as my assistant for the next six months.

John's arrival was fortuitous because just getting this many people organized was a surprisingly big job. We immediately started to work clearing trails and establishing new study sites throughout the Millaa Millaa area. Our intrepid travelers got a rapid introduction to leeches, stinging trees, and all the other nasty things we'd neglected to mention in our fliers. There was a wave of good-natured grumbling about "false advertisement" and "no one said anything about leeches" and so forth.

Most of the new arrivals were a bit nervous about working in the rainforest, so I gave them a minilecture about what to watch out for. The biggest concern, I explained, was getting lost. You couldn't see landmarks while in the dense forest, and because the understory was like a complex obstacle course, it was impossible to walk anywhere in a straight line. Rattans, vines, and lianas of every description snaked down from the canopy and often formed impenetrable masses on the ground. Trees of every size sprang up in dense profusion. Even after we'd cleared a rough trail, one still had to duck and bob through a maze of vines and buttresses. After a day in the rainforest even the fittest people discovered obscure muscles they'd long forgotten existed.

Getting lost was easy. In 1984 I'd spent a cold, wet night alone in a rainforest fragment after getting hopelessly disoriented while spotlighting. Far worse was the case of an American college student who'd decided to climb Queensland's highest mountain, Mt. Bartle Frere, alone. He disappeared into the rainforest and was never heard from again. Even a search by a big team of volunteers, led by his distraught parents, failed to discover any trace of his body.

These stories had the desired effect. Everyone now understood it was essential to carry a compass, and that they should stick to our marked trails whenever they could.

Infections were also a worry, I explained. Because even minor scratches can become badly infected, it was vital to treat cuts and bites with antibiotic ointment until they healed. The mammals we studied could also carry a number of diseases in their urine and saliva. The worst was leptospirosis, similar to malaria. For this reason anyone handling mammals or traps was required to wear rubber gloves and wash afterward with disinfectant. Several of the local biologists had contracted leptospirosis, which attacks the liver and other organs. I'd seen one fellow, Les Moore, nearly pass out after drinking a glass of wine, inadvertently bringing on a sudden attack of the disease.

The other thing everyone wanted to know about was snakes. Australia has the world's most dangerous snake fauna, with fully half of the species being venomous. The poisonous species are all in the cobra family—the Elapidae. Like cobras, most elapids are fast, alert snakes that rear up dramatically when threatened or curious. Some species even puff up their necks like cobras, though none have the cobra's hood or false-eye markings.

Aside from their remarkable speed, the thing that makes elapids so dangerous is their venom, a powerful neurotoxin that attacks the central nervous system. Death is normally by suffocation as the victim's nervous system is slowly overwhelmed. Our travelers were transfixed as I related this information, and I found myself perversely warming to my tales as I sensed their rapt attention. They all jolted when I suddenly lashed out at their faces, mimicking the strike of the death adder. Their eyes bulged as I described the fearsome taipan, a snake of such legendary speed and aggressiveness that it sometimes strikes its victim a half dozen times before he can run away. Their concern grew even greater when I offhandedly mentioned that a ten-foot taipan had slithered onto the main street of Atherton just the previous month.

But not to worry, I said, snakes here are rarely a problem so long as you don't startle or corner them. The key is to stomp your feet as you walk through the forest. If they feel you coming they'll just get out of your way. Our intrepid travelers obviously took this advice to heart, for the next morning they resembled a herd of angry buffaloes as they stomped and pounded their way through the forest. I suspect that every snake and animal for a ten-mile radius was at that moment fleeing hysterically. I finally had to explain that snakes were actually very sensitive to vibrations and would no doubt sense a group of our size from a great distance without all the extra foot-energy. They eyed me suspiciously when I said this—as though I were now changing my story just because I wanted to see a few animals in the bush.

The sad day of Christine's departure arrived. I drove her down to Cairns, lingering at the bus station until she boarded. She would be riding down to Brisbane, twelve hundred miles south, then flying to New Zealand to continue her travels. I didn't know how to thank her enough. As she boarded the bus she started to weep. Maybe, she said, she'd return in a few months if she could get another Australian visa.

On the drive home I thought about how close we'd become. We'd never really talked about our feelings but I knew I'd made a friend for life—she was that kind of person. I was already missing her.

My somber mood faded when I arrived home to a raucous scene. Our gang and an assortment of neighbor kids were engaged in a wild game of cricket in our backyard. Most of our travelers had no idea what they were doing. Rich, the American, was swinging an ersatz cricket bat like he was going to blast a home run. John, beer in hand, was yelling rudely in an effort to distract him. Gail was vigorously hurling a tennis ball at Rich. The neighbor kids, all of whom grew up playing cricket with its elaborate code of British civility, were laughing helplessly at the debacle.

John had quickly established himself as the group's wildest comedian. Nothing was too outrageous. Though I hadn't known him well at Berkeley, he'd responded to my campus ad for a field assistant and I'd immediately warmed to his laid-back manner. Blond and rather good looking, John, I quickly learned, was a raving extrovert, a consummate people person. That was an advantage now, because he and I were living and working with eight others. Being quieter myself, I was delighted to have an ad hoc social coordinator.

As might be imagined, the sudden influx of international travelers generated intense curiosity from the denizens of sleepy Millaa Millaa. The town was abuzz with news about us. The Millaa telegraph service—gossip at the speed of light—was functioning as well as ever. One day we'd had a flat tire south of town, which we'd quickly repaired. Entering the post office only minutes later I was amazed when the postmistress remarked on our flat tire, a performance repeated moments later at the takeaway. Incredible. What did these folks do for entertainment before we got here?

We had other impacts on Millaa's social dynamics—some not so good. One afternoon our gang strolled over to the pub while I stayed home to catch up on field notes, planning to join them an hour later. As I entered the pub I heard a great howl, which turned out to be Rich yelling because he'd missed a pool shot. The rest of our mob was carrying on raucously while the locals were all clustered at the opposite end of the pub, looking displaced and resentful.

Seeing the obvious, I herded my group out the door, ignoring their boozy protests. I explained firmly that we *had* to be more circumspect here; the locals tolerated us but would get far more annoyed if we acted like a bunch of wild tourists. They accepted this pretty well, I thought, and afterward their pub demeanor was better.

But the culture clash continued. John approached me a few days later, saying one of our neighbors had just chewed him out. What happened? I asked, alarm bells immediately ringing.

Well, John said, he was simply walking down to the takeaway when this old guy stormed out of his house and yelled at him to "Dress properly!"

My alarm bells became wailing sirens. And what, I asked, had he been wearing at the time? To which he offhandedly replied, "A sarong."

I groaned. I'd seen his sarong, a silky little thing of garish red and yellow. It would have drawn second looks in San Francisco.

"Were you wearing anything else?" I asked hopefully, but John shook his head.

"Let me get this straight," I said. "You sauntered down the main street of

Will Chaffee, John Vollmar (with headscarf), and me (with machete) at the end of a trapping session. The cage traps were for catching giant rats and marsupials. The horse belongs to a farmer who'd stopped by for a chat. PHOTO BY STEVE COMPORT.

Millaa Millaa buck naked but for a wispy little *sarong*? I don't suppose it occurred to you that the loggers and farmers around here don't see a lot of men in *skirts*?"

John just laughed and said he didn't see what the big deal was. Though I was sure the incident was reverberating throughout the township even as we spoke, John, I was quickly learning, wasn't one to sweat the details.

After a couple of weeks we settled into a routine of sorts. One of the hardest things about working with a mob this big was simply getting everyone awake, dressed, fed, and into the field by 8 A.M. or so. Tully had become an unexpected asset in this regard. He'd bolt into the house each morning in a blur of excitement, running in circles, wildly barking. I'd simply open a bedroom door and slam it shut as he rushed inside. Piercing screams followed as Tully leapt into beds and tried to lick everyone's face simultaneously. He'd be turfed out but would continue his rounds through the house until everyone was up and moving about. Far better than an alarm clock.

Our travelers were all good people but they were on holiday, looking for high adventure. I kept finding myself in a quasi-parental role, an unwelcome state of affairs as far as I was concerned. I'd been hoping John would develop into more of a taskmaster but that just wasn't his personality. He was a lovable rebel, not an authority figure. He was also a consummate night owl, sipping beers and strumming his guitar until 3 or 4 A.M., and chatting animatedly with whoever felt like accompanying him. At first I figured he was a big boy and could manage his own sleeping patterns, but I started to change my mind when he and his nocturnal cronies became so sleep-deprived they could barely keep their eyes open during the day.

Each morning the ten of us crammed into our little truck and headed off to the rainforest. To accommodate everybody we had built bench seats in the back, which we'd covered with a canvas shell. There was talk about painting a logo on the truck—a possum or tree-kangaroo—but we decided that with our lack of artistic ability it was probably best left alone.

About this time a ritual of sorts developed which we used to herald a new trapping session. Termed the "Inspection of the Troops," it entailed lining up our motley crew in front of the truck while I assessed their readiness. John would don severely bent glasses and adopt a profoundly idiotic expression. Rich would twist his bill-cap sideways and wave a syringe around like a drug addict teetering on his last legs. Dale might then emerge from the house proudly wearing Cathy's bra, while Cathy had piled bananas and mangoes upon her head like Lola Falana. The rest of the gang joined in similar spirit. They would puff out their chests and salute officiously, often facing backward.

My job was to march up and down, scowling deeply and waving my swagger stick—an old TV antenna. My gruff speeches paraphrased the recruiting slogan of the U.S. Marines: "Here at the Rainforest Fragmentation Project we're looking for a few good men—and women," I'd bellow ironically. Then I'd stare in dismay at our bumbling band of misfits. "But unfortunately all we could find was you lot, so I guess we'll just have to make do." I'd bark further insults until we couldn't keep a straight face anymore. Then we'd pile into the truck and head off for another week with the mammals.

After spending three weeks with us, Joachim, our German traveler, became convinced he was going to die. This happened on his first night of spotlighting.

Joachim was obviously high-strung. Earlier in the night John and I had laughed because he'd practically jumped out of his skin after coming upon a long-nosed bandicoot. Bandicoots are cat-sized marsupials that have the annoying habit of waiting until you're virtually on top of them then exploding out of the grass with a piercing squeak.

We'd hiked a couple of miles through the rainforest and eventually entered a clearing overgrown with shrubs and sasparilla trees. Hiding in one tree was a tree-kangaroo, but it was obscured by foliage. I asked Joachim to work his way through the shrubs and slap the tree so the kangaroo would shift into a better position. He looked unhappy but began shoving his way through the thick vegetation. As expected, the tree-kangaroo leapt down and loudly bolted off as Joachim approached.

Joachim returned saying that a spider had bitten him. I glanced at a tiny red mark on his hand and replied with a completely deadpan expression that he shouldn't worry, less than half of the local spider species were deadly. To me this was an obvious joke; there was only one seriously poisonous spider in north Queensland—the redback, a cousin of the black widow—and it didn't live in the rainforest. I winked at John and we continued our spotlighting without giving it a second thought.

A few minutes later Joachim turned to me and said that we had to go home *right now*. The color had completely drained from his face. When I asked what was wrong he clutched at his chest and said, "Bill, my heart!" I felt his pulse and it was thundering at an incredible rate. I was beginning to think I'd been too glib about the spider—Joachim truly looked awful. We quickly reversed our course. As we walked, John and I kept glancing nervously at Joachim, half expecting him to keel over at any moment.

After a mile or so Joachim let out a terrified yell. John and I spun around in alarm, but as we stared at him his expression slowly changed from one of panic to relief, and finally to sheepishness. It turned out that a cricket had jumped up and landed on his arm—that's why he'd yelled. Based on his overreaction, Joachim realized he was being utterly paranoid about the spider bite. His heart rate quickly dropped and he began laughing with shaky relief. John and I started laughing too. Soon the whole thing seemed hysterically funny; we could barely keep ourselves from falling over. We eventually dried our eyes, then backtracked to finish the census.

This incident, of course, became a long-running joke. Whenever Joachim was tired he'd clutch his chest and moan, "Bill, my heart!" It was good for him to joke about it. Otherwise, we would've teased him mercilessly.

I'd worked with other biologists previously but this was the first time I'd ever operated with travelers. In many ways the travelers compared favorably.

They were good workers and tended to do exactly what you asked them. This was in stark contrast to some biologists who wanted to hear elaborate scientific justifications for every decision, and had strong ideas of their own about how to do things. Obviously the travelers needed to be supervised, but that wasn't a problem with John and me both present.

At times I was really proud of our group. On several occasions we'd be zipping across a paddock and would bump into the farmer who owned the land we were working on. In rural Australia, an unwritten code under these circumstances is to stop for a "gasbag"—a leisurely talk. So I'd climb out of the truck to chat with the farmer about his cattle, the weather, the animals we were catching—but avoiding sensitive topics like politics and rainforest conservation. Sometimes we'd gasbag for a half hour or more. Though our crew was as antsy to get moving as I was, they always sat quietly in the truck, paragons of good manners. They could see how important it was to get along with the locals.

Once I emerged from the forest to find a wiry old man in his seventies furiously screaming at our crew. His repertoire of four-letter words—and his ability to string them together into creative combinations—was truly impressive. It turns out that we'd inadvertently trespassed on his land, which abutted the state forest. I learned later that my crew had stood there mutely for at least five minutes while the old guy—a retired logger—had vigorously abused them. Eventually he calmed down, looking almost purged, and once we explained what we were doing he became remarkably friendly, even offering us access to his land anytime we wanted. It was a victory of rural diplomacy.

After several weeks with our volunteers I was also pleased with our safety record. Aside from the inevitable bites and scratches, we'd had no significant illnesses or injuries, which was a big relief. This record came to a screeching halt, however, the day our gang decided to emulate Tarzan.

It started spontaneously. We'd just finished a day's trapping in a small forest fragment when John grabbed a liana and took off on a running leap, swinging out over a big clump of wait-a-whiles. As he swung back he cleared the thorny vines easily and landed gracefully on his feet. Not to be outdone, Rich took a turn and also cleared the vines. Dale followed next, but as he swung back a vine latched on to his arm, ripping his skin in a dozen places. I should have stopped the contest at this point, but foolishly let things carry on. Injuries accumulated quickly. Joachim got raked across the face by a vine and bled profusely, while Gail fell and sprained her ankle. The contest finally ended when Cathy plummeted right into the heart of the wait-a-while patch, sustaining myriad cuts and a badly bruised knee.

Joey MacMillan displaying the latest in rainforest fashions. Her stylish vest contains eartags, mammal scales, and other trapping gear. PHOTO BY STEVE COMPORT.

Though tough as nails, she had tears in her eyes, and we had to carry her out of the forest.

I cursed myself as my bruised and bleeding crew climbed slowly into the truck. In a quarter of an hour we'd sustained more injuries than we'd had in the past two months.

rainforest politics

While our volunteers were invaluable, one thing most of them failed to appreciate was the sanctity of the data. In a scientific research project the data are all-important—the raison d'être. Extreme care needs to be taken to ensure that the data are valid and accurate. Uncertainties are dealt with harshly: a common maxim in science is, if you're not utterly certain about the validity of the data, throw them away. This is because, like a rotten apple, one bad data point can bias an entire investigation.

Because most of my crew lacked scientific training, I carped endlessly about data quality in an effort to instill a proper sense of reverence and care. A good scientist frets about his data, I explained, and always stores copies of important notebooks in several different locations to ensure against an unforeseen catastrophe. Universities are rife with stories of near-suicidal graduate students who lost three years of data in a computer crash, or had a critical notebook fall overboard while boating up the Amazon.

While they endured my proselytizing, I could tell some of our gang found my paranoia about data quality a little extreme. "He's a bit over the top about the data, isn't he?" I overheard Cathy remark one day. Cathy recorded the data in the field and carried the all-important notebook. Her boyfriend, Dale, joked about the evil things that would happen to her if she ever lost the notebook. "If you should find yourself falling off a cliff, just be sure to throw Bill the notebook before you disappear," he quipped. "Otherwise you'll be in *real* trouble."

During one of our trapping sessions the gang conspired to see how I'd react to losing a big chunk of crucial data. We'd been working all week in a seventy-acre fragment located on a dairy farm. To reach the fragment we had

We often had to pass through herds of dairy cows on our way to the forest fragments each morning. PHOTO BY AUTHOR.

to drive by the farmer's milking shed, sometimes nudging through a herd of dairy cows as they patiently waited to be milked and fed. There were gates to be opened and closed on either side of the shed, a job everyone hated because that was where the waste from the milking shed accumulated. To reach the gates one of us had to get out of the truck and wade through a pungent slurry of urine, mud, and cow crap, deep green in color and the consistency of a milkshake.

As we approached the shed one morning John began to talk at length about how vital it was to look after the data. This seemed a little out of character at the time. Cathy jumped out of the truck to open the gates, conspicuously carrying the bright red data folder as she did so. The idea, I gather, was that Cathy was supposed to pretend to stumble, drop the notebook into the cow dung, then slip again and step on it, ensuring that it became completely drenched in the horrible stuff. I was supposed to freak out, and probably would have if the gag hadn't gone completely awry.

For one thing, John's less-than-subtle foreshadowing made me vaguely suspicious. I would have had to be blind, moreover, not to notice the flagrant way Cathy was waving the notebook around—just to be sure I saw it—as she tried to open the gate. But the real coup de grace was the complicated physical execution required of Cathy: as she pretended to stumble and then tried to step on the notebook, she slipped, lost her balance, and fell face-first into ten inches of pungent cow shit. Her pants, chest, and arms were completely drenched, and big droplets splashed her face and hair. The data folder—which was actually empty—floated on the surface a few feet away.

I jumped out of the truck in alarm—for Cathy, not for the data—then was surprised to hear everybody laughing hysterically. John and Dale had completely lost it and were barely able to stay on their feet. Cathy was looking sheepish and not too happy to be dipped in cow dung. We made her ride in the back of the truck until we could get home and hose her off thoroughly. I think the disastrous way the joke turned out was actually funnier than the original plan. And fortunately that was the last time anyone tried to tease me about losing the data.

Because I had such a big mob of helpers I decided we should begin our habitat work. This was a intensive job that involved measuring the rainforest itself: its architecture, topography, and soils, and the abundances of certain plant species such as lianas, wait-a-whiles, and stinging trees that were indicators of the health of the forest.

I had two reasons for doing this. First, I needed to understand which features of the forest were important to the mammals we were studying. For example, did certain species avoid disturbed forest, while others preferred areas along creek lines, and so forth? Second, I wanted to contrast the rainforest in small fragments with that in larger forest tracts. My subjective impression was that the fragments were not just small samples of intact forest, but were ecologically different. I wanted to know how they differed from normal rainforest—and why.

Our days suddenly became very long. We'd trap in the morning, have a quick lunch in the field, then do four or five hours of habitat work in the afternoon. After dinner, John and I often went spotlighting until midnight.

The habitat work was repetitive and boring—laying out quadrats, counting plants, measuring the sizes of trees, recording the elevation and slope, taking wide-angle photographs of the canopy. We now had fifty-two trapping grids scattered around the tableland and we recorded hundreds of measurements at each grid. If we worked quickly we could finish two grids in an afternoon.

Our travelers put in a spirited performance, but they were getting worn down by all the hours in the forest. I'd bribe them with cokes and candy bars, or turn the afternoon into a contest. "If we finish two grids by 4 P.M.," I'd promise, "I'll buy us a carton of beer." They'd whoop with joy and work with frantic intensity, never missing a beer-induced deadline. With that mob of travelers and a truckload of beer I reckon we could have built an Egyptian pyramid.

The habitat work wasn't just boring, it was occasionally risky to life and limb. One afternoon a huge centipede—easily eight inches long—ran up Dale's arm. He screamed and shook like he was possessed by the devil, and the centipede went flying. I was greatly relieved he wasn't bitten. Centipedes have impressive poison fangs for subduing their prey, which can include insects and even small birds and mammals. A biologist I know was bitten by a big centipede in north Queensland and her arm ballooned up to twice its normal size.

The worst injury was sustained by poor Patrick, the French Canadian. Some of our grids were located in such steep terrain that we needed ropes to reach them. We were in one such area, an abrupt hilltop in a fragment that had been savaged by the recent cyclone. Stinging trees and wait-a-whiles sprang up everywhere. One of our tasks was counting fallen fruits—potential mammal food—in meter-square quadrats laid out on the ground. The fruits were often obscured by fallen leaves, so we'd sweep our hands back and forth through the leaves to expose any hidden fruits.

Because the slope was so steep, I was having a hard time reaching my quadrat, so I asked Patrick, who was just above me, to check it for any fruits. Growing in the quadrat were dozens of tiny stinging trees which I'd assumed Patrick had seen. Before I could say anything he quickly swept his hands back and forth through the leaves and stingers. I yelled in alarm and he looked surprised. Then his eyes went wide as waves of burning pain began to pour over him.

I felt awful. Because they contain so many nerves, the hands are one of the worst places to get hit by stingers, and Patrick had just absorbed a huge dose of toxic spines. The color drained from his face as he stared at his hands, gritting his teeth and trying not to break down in tears. The lymph nodes in his armpits began to throb as his body tried to deal with the sudden influx of alkaloids.

Gently holding his upper arms, John and I helped him to the truck, then laid him down in the back. The rest of the group jumped in and we headed straight for the ambulance center in Millaa Millaa.

By the time we reached Millaa, Patrick had regained a little color. The initial pain had begun to subside. We walked into the ambulance center and I explained to two laconic-looking chaps in uniforms what had happened.

Though I knew there was no truly effective treatment for stinging tree, I was hoping they'd at least give Patrick some painkillers or anaesthetic ointment. But the ambulance officer said the only thing he could do was spray on "liquid skin," which dries to form a tough layer over the skin surface. When the layer eventually peels off, he said, it would extract a lot of the poison spines with it.

We figured he knew what he was doing, so Patrick told him to go ahead. The officer held up a spray can and shot of blast of icy-cold liquid onto Patrick's hand.

Patrick stood up and screamed. The intense cold from the spray had made his skin contract, breaking the spines and releasing more poison. Unbelievably, the ambulance officer started to laugh.

At this point I should mention that, although sadistic, laughing at someone in pain is quite consistent with the Australian sense of humor—so long as they're not actually dying. Patrick wasn't dying, although he probably felt like he was at this moment. The ambulance officer shot another blast of freezing spray onto his skin and Patrick bellowed again.

This time the other officer started laughing too. They just couldn't stop themselves. Patrick was howling in pain every time they sprayed him and they just kept convulsing in laughter. Then we started to laugh too. We tried not to, but we simply couldn't help it. Patrick was screaming again and again and the rest of us were laughing our heads off. The more murderous his expression became the funnier it seemed.

After a few more minutes of torture we finally got the poor bastard out of there. He'd turned pale again so we took him home and fed him several shots of whiskey. Then he lay down and tried to sleep, his hands carefully propped up so as not to touch anything.

John and I sat at the kitchen table mulling over the incident. "Just promise me," he said solemnly, "If I ever get hurt—take me to Atherton, to Malanda, to Cairns, anywhere. Just don't take me to those sadistic bastards at the Millaa Millaa ambulance center."

Despite such small disasters, our efforts in the field were beginning to pay off. At the end of each week we uploaded the habitat data into my computer, trying not to work at milking time—dawn or late afternoon—when the electricity plummeted and the computer groaned and fluttered.

The patterns on the computer screen confirmed my initial suspicions. Something important was happening to the forest fragments. Many were massively disturbed—above and beyond that caused elsewhere by the cyclone. Countless trees had been knocked down, making the forest's canopy look like Swiss cheese. Normally, the continuous canopy is the protective skin of the forest. There is little sunlight on the forest floor, and temperatures remain cool and stable. The canopy traps moisture, maintaining high humidity, and there is little wind. Because rainforests are ancient ecosystems, myriad varieties of plants and animals have become specialized for living under these cool, moist, dark conditions.

The disruption of the canopy in fragments leads to an array of ecological changes. Because light levels rise sharply, weedy plants proliferate in the understory—stinging trees, wait-a-whiles, and even exotic species like lantana and potato vine, a relative of tobacco. Lianas—woody vines—love disturbed forest, and proliferate in dense tangles. The forest becomes drier and temperatures fluctuate more widely. Broken trees and limbs litter the forest floor. The complex architecture of the forest is profoundly altered.

Why were the fragments so badly disturbed? I thought I could guess the answer. They were highly vulnerable to winds. In addition to the occasional cyclone, Australian rainforests are buffeted by strong winds. In a large forest tract, only the exposed mountaintops are badly damaged. But the fragments are different. Encircled by a barren landscape, they are battered by windstorms. As a consequence, fragmentation wasn't affecting just the animals. Many aspects of the rainforest—its architecture, composition, dynamics, and microclimate—were also being altered.

At about this time I was asked by the Rainforest Conservation Society of Cairns to give a talk at their annual meeting. Our project was becoming quite well known in the area, and I'd just published an article in an Australian magazine that highlighted many of the ecological risks of logging rainforests. I also sat on the conservation committee of the American Society of Mammalogists, an international scientific organization, and had sponsored a resolution the previous year that lobbied the Queensland government to halt rainforest logging. The resolution focused on many of the technical arguments against logging and received widespread media coverage in Australia.

Rainforest logging in Queensland was becoming an increasingly contentious issue. Many Australians were aware of the devastating pace of rainforest destruction in developing countries, and many were appalled that Australia, a developed country, could do little better. Increasingly, attention was being focused on north Queensland, where an intransigent state government, led by its archconservative premier, Joh Bjelke-Petersen, steadfastly refused to curtail logging.

To understand what is happening to the world's tropical forests, it's important to draw a distinction between logging and forest clearing. Both are big problems, but they're very different processes. The world is currently losing outright about forty million acres of tropical forest each year—much of it cleared for cattle pastures or slash-and-burn farming. Another fifteen million acres are being selectively logged, mostly virgin forest. Selective logging doesn't destroy the forest, but it causes major disturbances.

In a typical logging operation, bulldozers and other heavy equipment are used to extract timber from the forest. Although harvests are modest—typically three to ten trees per acre—a third or more of the trees may be killed by the labyrinth of roads and loading ramps created to drag logs from the forest. Up to half of the canopy cover is removed. Soil disturbances can be extensive, leading to erosion and sedimentation of streams. The forest's structure and microclimate are profoundly altered, at least for the first few decades after logging.

But in many developing countries where population pressure is intense, the indirect effects of logging are worse than the logging itself. Hunters use the newly created roads to penetrate deep into logged forests. The loggers themselves are often major hunters; in northern Borneo, for example, a single large logging camp was estimated to consume over seventy thousand pounds of wildlife meat each year.

Logging also promotes an influx of slash-and-burn farmers, who clear and burn the forest for crops. Because rainforest soils are quickly depleted after burning, the farmers must continually clear new forest areas. The final result is a mosaic of degraded, scrubby vegetation that bears little resemblance to primary rainforest. In areas as diverse as Borneo, New Guinea, and the Amazon, I have watched as slash-and-burn farmers follow almost on the heels of loggers.

In north Queensland, the logging industry was very well entrenched, but was experiencing a slow death. It was largely a matter of past overcutting. The industry had grown too large and had overcut the forests, much in the way that loggers overharvested the coniferous forests of the Pacific Northwest in the United States. Although harvest levels had been lowered and then lowered again, the Queensland industry was still dependent on logging virgin forest to remain viable.

This was where my hackles started to rise. The tropical rainforests of Australia are tiny in extent—only 0.2 percent of the continent's land area—yet have been massively exploited over the last century. Most rainforests on productive soils were destroyed for agriculture, and much of the remainder fragmented or logged. These forests—relics of some of the most ancient ecosystems on the planet—were being subjected to a powerful one-two punch: clearing and logging. The area of undisturbed forest was rapidly diminishing.

Australians were becoming increasingly alarmed about the management of Queensland's rainforests. Bjelke-Petersen's decision to punch a forty-mile road through the famous Daintree rainforest in 1984 had provoked a bitter national controversy. Nature-loving hippies and other protestors had tried to blockade the road, chaining themselves to trees or burying themselves in the ground. They were systematically dug out and arrested.

While I personally didn't support the protestors' tactics, I could understand them. For a democracy, Queensland in the 1980s was surprisingly undemocratic. There were few public forums or opportunities for residents to oppose government decisions. Bjelke-Petersen—once a rural farmer—had been in power for over twenty years and his style of governance was remarkably arrogant. "Don't you worry about that!" was his famous rejoinder whenever he received a question he didn't like, which he'd simply refuse to answer. The premier held strong views on many topics. He didn't like homosexuals, intellectuals, or conservationists—"greenies" in the Australian vernacular. He was 100 percent probusiness and some of his closest associates were land developers. Nobody had neutral views on Joh Bjelke-Petersen—you either loved him or hated him.

Partly because of Bjelke-Petersen, Queensland was becoming increasingly polarized by conservation issues. The sunbelt state was growing rapidly, and development controversies filled the newspapers almost daily. The Daintree road had helped to bring the plight of north Queensland's rainforests into sharp national focus. Some local residents were infuriated by the confrontational tactics of the protestors, while others were appalled by the premier's unilateral decision to bulldoze an unsurveyed road through the nation's last major tract of lowland rainforest. Peter Stanton, the highly respected regional director of the Queensland National Parks and Wildlife Service—and technically an employee of Bjelke-Petersen—had come out in public opposition to the premier, and many others had followed his example.

But on the Atherton Tableland things were different. This was the heart of logging country, just like the rural timber towns of Oregon and Washington. Most towns had timber mills and were populated by loggers and their families. Tough and fiercely independent, they held no ambivalence about logging or other rainforest conservation issues. To be labeled a greenie was the kiss of death.

For this reason I was a little nervous about my talk before the Rainforest Conservation Society, which was reported by the local newspapers. I depicted the past destruction of forests in the region and how some of the rarest rainforest types were now being intensively logged. I also described our research on the tableland, illustrating my talk with slides of beautiful rainforest animals. Our project, I explained, was revealing that fragmentation causes a progressive decay of the complex rainforest ecosystem. The talk went well, and the audience peppered me with questions when I finished.

Back in Millaa Millaa, I was a little surprised that no one mentioned my talk in Cairns. I assumed it had escaped everyone's attention. I would later realize just how naive I was to think so.

gone troppo

Gone troppo: slang, Australian. A form of temporary insanity occasionally occurring among settlers of European extraction. Often associated with protracted heat waves at tropical latitudes. Example: "He's gone completely troppo and has been talking gibberish all week."

"Hey Bill," Rich yelled from the kitchen, "the Millaa cop wants to talk to you."

John and I were in the midst of a serious game of Ping-Pong, the final in a best-of-seven-game series. We had a six-pack of beer riding on this latest competition. The Ping-Pong table and a small television were intended to help keep the crew occupied when we weren't working.

Damn. "What does he want?" I yelled back.

"He didn't say, but he doesn't look happy."

He never looked happy. Drenched in sweat, I pulled on a tank top and walked outside.

"How ya goin', Mark?" I asked. He and I were now on a first-name basis of sorts. I'd gone out of my way to say g'day whenever I happened to see him.

"I've just come to give you a warning." He said. "If things keep carrying on like they have been, you're going to get searched." He didn't have to say for what. He was talking about drugs—marijuana.

"Listen, Mark," I said, looking him straight in the eye, "I've told you before. I give these guys a big speech when they arrive here. Strictly *no* drugs. I tell them that if they're carrying any pot to get it the hell off my property. As far as I can tell they've all taken this seriously."

"Well, I've had reports of people—*your* people—hanging out over there behind the church," he countered, pointing at the small wooden church across the street. "That's pretty suspicious behavior if you ask me."

Dammit. It *did* sound suspicious. "Uh, look," I said, "I'll check into it. Whatever's going on I'll put a stop to it. Seriously, Mark, I don't want any trouble around here. Especially with the law. We're here to do our fieldwork and that's it. Really."

"Okay," he said gravely, starting the car. "Just be sure it *does* stop. For good."

Not a pleasant experience, though I actually appreciated the warning—far better than getting searched, the stigma of which would be awful in a little town like Millaa Millaa, even if they didn't find a thing.

I walked back into the house feeling seriously pissed off. Everyone was looking at me expectantly. "Let's have a meeting," I said.

Clustered together in the living room, I made the most forceful speech I'd ever delivered to them. Several sets of eyes dropped to the floor. We could lose *everything* we were working for here as a result of one person's stupidity. Keeping marijuana off my property did *not* mean sneaking across the street and getting stoned behind the church! I ended with a plea for their cooperation, and repeated that anyone who couldn't abide by the no-drug rule would have to leave immediately.

Everyone was silent, so I walked outside for a cigarette. I was staring off at the horizon when two of our travelers approached me contritely and promised I wouldn't have to worry about drugs anymore.

The day was quickly approaching when several volunteers—Dale, Cathy, Rich, and Joachim—needed to move on. There was certainly no shortage of people to take their places. We'd had to turn down many eager travelers, some of whom phoned back repeatedly to see if we'd had any openings. Each volunteer was required to stay at least one month; otherwise we'd lack any kind of cohesiveness.

As their departure approached, Dale grew increasingly moody. At times he seemed utterly preoccupied. One day he did such a terrible job of setting traps that I had to talk to him about it. On another day he screamed wrathfully at me for driving too fast. When he'd first arrived he'd been a fun and contented person, so the change did not go unnoticed.

Finally, Dale confided in us: he and Cathy were on the verge of splitting up. This was a great shock—they'd seemed so happy together. But while Dale was head over heels in love, Cathy had decided she just wasn't ready for a permanent relationship. They'd be going separate ways on their touring bikes when they left here. We could all see why he was crazy about her. She was a bright, attractive, and incredibly together person. A South African, she'd left the country in the early 1980s because she could no longer tolerate living under the nation's discriminatory apartheid regime.

The week before our four friends departed was hell. On top of the usual trapping and spotlighting, we were pushing hard to finish the habitat work. We'd had a week of drenching rain, working in a steep, massively disturbed fragment—a quagmire of thorns and mud. Nerves were stretched to the limit. To exhort the troops to finish I'd promised not only copious amounts of beer but whiskey and a big barbecue.

We dragged ourselves home at the end of the week, cranky and hollow-eyed with exhaustion. We were out of food, so part of the gang drove fifteen miles to Malanda, to replenish supplies. Meanwhile, Rich and I lugged three cases of beer and two bottles of whiskey from the local pub. It was a lot of booze, but we'd earned it. We were in need of a big blowout, not only to send our mates off in style but to vent the stress that had been building up volcanically during the week.

The shopping crew arrived home with a huge load of food, and we went to work preparing our feast—barbecued steaks and chicken, fried onions and potatoes, salads, and a cake for dessert. With the exception of Dale and Cathy, who retired quietly for a final night together, we ate our fill and then settled in for some serious boozing. After a couple hours of guzzling beer we started on the whiskey, playing raucous drinking games until we were wobbly on our feet. Amazingly, John staggered down to the pub and returned with two more bottles of whiskey. The party got wilder. We were laughing, screaming, dancing—our heads swimming with the exuberance of being dry and comfortable and out of that damned, leech-infested rainforest. The night faded into a blur, then exploded.

John had been irreverently comparing Gail to a brush turkey, an odd Australian bird that buries its eggs to be incubated in big mounds of leaves, just like reptiles. Brush turkeys have distinctive red-and-yellow wattles around their necks. John threw another verbal barb at Gail—my booze-addled brain didn't catch it—and she screamed in indignation. "I'll make you look like a brush turkey," she said, grabbing a bottle of ketchup and dumping it on John's head.

We all roared. "Wait!" Rich yelled, "he needs some yellow too." He grabbed a squeeze bottle of mustard and squirted a design on John's head. Now John *did* vaguely resemble a brush turkey. Not to be outdone, John opened the refrigerator and emerged with an egg, which he ground into Gail's hair.

John Vollmar at the beginning of our legendary food fight. PHOTO BY AUTHOR.

Thus began one of the greatest food fights in Australian history. Having just purchased three hundred dollars' worth of food, we were exceptionally well armed. Food flew everywhere. Coalitions formed briefly like tag-team wrestling, then collapsed from treachery. Eggs were pitched like baseballs, bags of flour dumped like bombs onto unsuspecting heads, peanut butter smeared across faces. Pitchers of water and milk and orange juice got thrown for good measure. We even dug into our stock of mammal bait, spraying each other with bottles of vanilla essence, its cloying smell wafting through the house. We screamed like wild banshees as we raced from one end of the house to the other, pelting each other with dill pickles and squirting maple syrup.

The lunatics had escaped from the asylum. After endless hours in the muddy, thorn-ridden rainforest, we had gone completely, utterly troppo.

Dale finally emerged from his bedroom, his face flushed and furious. His jaw dropped as he surveyed the house. The kitchen was the worst hit. It had gallons of food and liquid of every sort dripping from the walls. The floor was completely cloaked in a brownish morass of flour, juice, and a dozen other substances. We food-warriors were splashed with colorful concoctions like psychedelic war paint.

Dale shouted at us for a few moments then returned to his room and his final night with Cathy. Even in our drunken stupor we were a little taken aback. Dale thought we were being incredibly insensitive under the circumstances, but we hadn't planned for culinary Armageddon. It just happened.

The house was a catastrophe and our clothes were literally stuck to our bodies. We decided to drive to Millaa Millaa Falls to wash off in the swimming hole. I vaguely thought it was a dumb thing to do, but at least there would be no traffic on the road at that hour. We pulled into the public parking lot next to the swimming hole and the gang dashed off to dive in. I just plopped down on the pavement next to the truck, not quite ready for frigid water.

I'm not sure how long I had been sitting there when a car approached. Completely drenched in a colorful smorgasbord, I looked like an apparition from outer space. I was contemplating diving for cover when the car zipped into the parking lot and swung in next to me.

It was Mark, the Millaa Millaa cop.

He shone his spotlight on me for a long moment, then swung the light toward the swimming hole. Then the spotlight veered back to me. He extinguished the light and just stared at me, shaking his head. I'm sure every suspicion he'd ever held about our degenerate lifestyle had just been confirmed—and then some.

What the hell. "At least we're not smoking pot," I said defiantly.

He didn't say a word, just started his car and drove off. I was stunned; I was sure we were all going to jail. I staggered down to the swimming hole and jumped in.

I awoke the next morning to face the Mother of All Hangovers—my head gonged and my stomach bungee-jumped—and I still had leftovers in my hair and other obscure places. I wasn't ready to face reality, but reality insisted on rearing its ugly head.

In daylight the house looked even worse than it had the night before. The swamp that coated the kitchen floor and walls had congealed into a thick brownish sludge. Dale and Cathy were nowhere to be seen. Their touring bicycles were gone; they'd left without saying goodbye. Cathy had left a little note of thanks, but there was nothing from Dale.

Dammit. I shook John awake and told him we had to find Dale. We knew he was planning to head northward toward Atherton, while Cathy was going south. We stumbled out to the truck and raced off. After several miles we saw Dale peddling along the road. We were pretty contrite, and he finally forgave us for spoiling his last night with Cathy. We'd been through a lot together and had formed a good bond. It would have been an awful shame to end it on such a bad note. We talked for a while then gave Dale a big hug and headed back to Millaa, greatly relieved we'd found him. He'd promised to stay in touch and let us know how he was doing.

As we walked back into the house one of the neighbor kids—a regular visitor—was gawking at the kitchen. "What . . . did somebody spew?!" he asked in a horrified tone. John and I dissolved into hysterics.

For the rest of that day we moved very slowly. I collapsed for a nap and awoke to find that John and Sue, the Englishwoman, had dragged most of the furniture outside and were hosing out the house. The rest of the gang were sleeping under trees in the backyard. I ambled out to join them, using Tully as a pillow.

The next day John and I drove Rich and Joachim down to Cairns, where the three of them embarked on a pub crawl that I just couldn't face. John and I agreed to meet the following day. I holed up at a hostel and awoke early the next morning for a boat trip out to the reef. It had been two years since I'd gone snorkeling on the Great Barrier Reef and I was keen to see it again.

The boat left the wharf at 8 A.M. and chugged out through the brilliant blue water of Cairns harbor. From their accents I could tell that most of the two dozen or so tourists on board were Australians.

In the outer harbor we passed an American navy ship inbound for Cairns, its sailors crowding the decks in smart dress uniforms. They were probably coming in for a furlough. I could sense their anticipation—maybe this was their first visit to Australia. From the ship a hundred pairs of eyes watched us chug by in our small tourist boat.

No one waved. Not a single person on our small boat tried to welcome the Americans. I couldn't believe it. I'm not a chest-thumping patriot, but my father was in the navy during the Korean War. I could imagine how he would've felt calling into the port of a so-called friendly nation and having the locals just stare at him.

For a moment I was disgusted with my fellow passengers. I thought about how quiet and undemonstrative Australians could be. Many of the silly game shows that crowded American television also occurred here, and it was hilarious to see how understated the show's participants were. While an American would be frothing at the mouth at the prospect of winning a car, an Aussie who'd just won a sparkling new Mercedes would just smile and shyly say, "Great." Trying to instill a sense of excitement, the hosts on those Australian game shows really earned their money.

The other thing I was seeing here was the ambivalence of Australians toward Americans. The United States is respected as a powerful ally and democracy, but many Aussies consider Americans ostentatious and loud—sometimes arrogant. Where Americans admire ambition and success, Australians respect humility and stoicism. Though superficially similar, there are actually many subtle and not-so-subtle differences between the Australian and American characters.

As these thoughts sped through my mind, the navy ship passed by us. I could imagine how put-off the sailors must be feeling. In Australia, of all places, they'd probably expected a friendly welcome.

The hell with it. I ran up to the observation deck and started waving frantically. A hundred arms waved back instantly. Faced with this withering demonstration of friendliness, the Australians began waving back to the Yanks. Suddenly everybody was waving to everybody.

The reef was as spectacular as always, a kaleidoscope of colors and shapes. It was a sunny day and the reef fish—red emperors, parrotfish, coral trout—flashed brilliantly as they darted among the corals. I snatched breaths of air then dove down to glide just above the reef, my face skimming above starfish and sea anemones and a hundred kinds of corals. Giant clams—some five feet wide—were scattered about. Their fleshy lips hold dozens of rudimentary eyes and are tinted emerald green, the result of symbiotic algae that live in the clam's flesh and provide energy from the sun.

Bommies—giant coral heads—rose up majestically from the sea floor thirty feet below. Black-tipped reef sharks swam in lazy circles. I dove down to the sandy bottom and came up with a two-foot sea cucumber, a bizarre animal distantly related to starfish, though it looked more like a giant phallus. The sea cucumber spewed water and slowly deflated as I held it above the water. I released it and watched it drift back down to the bottom.

Before I knew it our two hours on the reef were over. I climbed back onto the boat with the other tourists and settled into a deck chair for the return home. The steady drone of the boat motor soon lulled me to sleep.

John and I met at the hostel later that afternoon. I was tired and sunburned, but fully rejuvenated. It seemed like weeks since I'd been in the rainforest, not merely three days. We headed back up to Millaa Millaa. Like us, the rest of the gang had recovered from our party and were ready to go back to work.

Several days later Dale called to tell us he'd reached the Daintree River, forty miles north of Cairns. I could tell he was still in a turmoil over Cathy, and I reminded him he could come back and join us at any time—we were his surrogate family. No, he said, he was going to shoot for Cooktown, a remote town fifty miles further north. He'd phone us again when he could.

The next evening I got another call—from Cathy. Though normally cultured and confident, she now sounded mournful and sad. "Bill," she said, "I've made an awful mistake. I *need* Dale." She explained she hadn't been able to stop thinking about him—she couldn't eat, couldn't sleep. She'd realized now that he was the one for her. She *had* to find him. Would I help her?

I felt like whooping for joy. I was ecstatic for Dale—for both of them. I told her what I knew of Dale's latest whereabouts and promised I'd pass along her message the second I heard from him. I was busting to tell him.

But he didn't call. Every time the phone rang we all jumped up eagerly to grab it, but it was never Dale. After a few days I tried phoning several of the hostels where he might be staying—Cooktown, Mossman, Cairns. I left urgent messages for him to call me. But no luck. Cathy phoned back twice to see if I'd heard from him, but I had nothing to report.

I'd almost given up hope when two weeks later the phone rang. "Hey," Dale said, "how is everybody?" He sounded relaxed, far better than the last time we'd spoken.

"You're not going to believe this, pal," I said. "*Have I got news for you.*"

the higher-mammal crew

The alarm buzzed at 1 A.M. Half awake, I made a cup of coffee and forced down a few quick gulps before dialing a long series of numbers. After several rings the phone was answered by a secretary with a breezy California accent. I was making my monthly call to Bill Lidicker, my major professor.

It was 10 A.M. in Berkeley and Bill answered in his usual concise way. In his earlier days he'd been a veritable boy genius, completing his Ph.D. at the University of Illinois by the age of twenty-four. He immediately accepted a faculty position at UC Berkeley where one of his first graduate students, George Heinsohn, was older than he was.

Now in his early fifties, Bill had spent nearly three decades at Berkeley. By coincidence, after finishing his doctorate, George Heinsohn moved to Australia where he took a faculty position at James Cook University in north Queensland. When I first visited the region for a few months in 1984, George persuaded the equipment manager at the university to lend me a brand-new pickup and thousands of dollars' worth of mammal traps, spotlights, and camping equipment. I hadn't met George before then, but because of the Lidicker connection we felt like academic brethren.

As a major professor, Bill was pretty sparing with his advice—something I liked—but when he did offer suggestions they were usually excellent ideas. He was also a quick and critical thinker, so I tried to be on my toes when I phoned him.

We spent several minutes reviewing the project, and I reported that the fieldwork was going surprisingly well thanks to the efforts of our volunteers. I sketched recent events but left out a few minor details such as food fights, romantic dramas, and our latest run-ins with the Millaa cop. I also explained that rainforest conservation issues were really heating up. Australia's federal government was being pressured to declare the rainforests of north Queensland a World Heritage Area. If this happened it was almost certain that rainforest logging in the region would be stopped.

Over the years many people have asked me, Why Australia? Why go Down Under to study rainforest fragmentation when there are so many degraded landscapes to choose from in South or Central America? There are two answers to that question—one short and one long. The short answer is Bill Lidicker. The long answer is, well, long.

Even before I'd arrived at Berkeley as a rookie graduate student, I'd decided I wanted to study how the fragmentation of rainforests affects the animals that live within them. This quest had evolved for several reasons: it was a massively relevant issue from a conservation perspective; it was exotic and challenging; and I'd developed an interest in tropical fauna while working summers in zoos and wild animal parks in the United States and England. I had not, however, decided what kinds of animals I wanted to study or where I'd go to study them.

I initially thought birds would be a good group to work with, but Bill—a dyed-in-the-wool mammalogist—made it clear that while I was free to study birds, if I did so his interest in my project would be substantially reduced. I wasn't a complete idiot. On that day I made a career-changing decision: I became a mammalogist too—a decision, I must say, I've never regretted.

This left the question of where I would conduct my fieldwork. Latin America was the obvious choice but there was one small problem—my language skills. Like so many Americans I was raised to be appallingly monolingual. I'd picked up some rudimentary Spanish while traveling in Mexico, but I was having a hard time imagining myself setting up a complex field project in a remote area while trying to ask the locals how to buy ten-amp fuses or mammal bait using wild gesticulations and a hundred-word vocabulary. The prospect did not seem promising.

It was for this reason, perhaps, that Bill's suggestion of a visit to north Queensland fell on receptive ears. Up to that point I hadn't even realized that there *was* rainforest in Australia. The local residents even spoke English, sort of. The clincher was that Bill had himself studied in Australia years ago and was keen to get a student working there now, trying to maintain a vicarious involvement with the unique Australian mammals. He was positive there had been a lot of logging and forest clearing in north Queensland, and offered to put me in touch with some Australian colleagues like George Heinsohn who might be able to help me find a forest to study.

I wrote letters to a half dozen people and virtually everyone said the same thing: the Atherton Tableland was the ideal place to study rainforest fragmentation. Not only was the area of extreme biological importance, but many of the fragments had been isolated over half a century earlier. This was a key feature. The process of biological decay takes time. Populations wither and die slowly, and can take decades or longer to disappear completely. In much of the developing world it is hard to find fragments more than ten or twenty years old, because in most areas large-scale deforestation has occurred only recently. But the fragments of the Atherton Tableland were old enough for many consequences of fragmentation to become manifest. Like a looking glass into the future, these remnants might provide insights into the eventual fate of degraded rainforests elsewhere in the world.

I finished my report to Bill and jotted down several suggestions he'd made. As I hung up, I thought about him sitting in his office. A voracious reader, each month he received dozens of journals and wildlife magazines, which over the years had grown into massive stacks that towered like lopsided skyscrapers from every horizontal surface in his office. I used to kid Bill that when the inevitable big earthquake finally hit the Bay Area, they'd never find his body beneath his thousands of books and journals.

Following the departure of Dale, Cathy, and the others we needed some fresh volunteers to bring us back up to full strength. I drove down to Cairns to pick up the latest recruits: Will, a lively eighteen-year-old from Boston; Mike, a suave Canadian; and two Brits, Bridgette and Martin.

During the drive home Bridgette and Martin sat in the back while I chatted up front with Will and Mike. Will was enthusiastic and wide-eyed at the thought of working in the rainforest. I gave him and Mike the usual spiel, explaining what we were doing and trying to impress upon them the serious nature of our work. The tale of our wild food fight had spread like a cyclone through Millaa Millaa, and since then I'd been making a concerted effort to inject some sobriety into our operation.

As we wound up the steep mountain road we passed some road construction, and I had to shift to the opposite lane for a few hundred yards. I was a little tired and, after clearing the construction, simply forgot to move back to the other side of the road. Chatting away amiably, none of us three North Americans noticed that we were driving on the right-hand side—the wrong side Down Under. I was in the midst of explaining to Will and Mike how important it was for us to have a professional field operation when a huge freight truck suddenly came roaring around a corner straight at us.

We all screamed. I jerked the wheel to the left while the trucker slammed on his brakes and veered right. Our tires screeched shrilly and we slid out of control toward the road verge, which dropped off almost vertically for several hundred feet. The pickup tight-roped the precipice between the road and oblivion, finally lurching to a stop only inches from the edge. The trucker, stopped fifty yards away, fired angry blasts at us on his air-horn.

Martin and Bridgette were picking themselves up off the floor of the pickup. Mike had opened the door and crawled out onto the ground, looking like he was about to bestow a papal kiss to the gravel beneath him. Will's eyes were bulging fearfully as he stared at me. I could see him thinking, *You call this a professional operation??*

I gulped deep breaths and apologized profusely to everyone. We sort of laughed it off, but as we drove on, I could sense an edge of nervousness in my companions that hadn't been there before.

Will, I discovered, had a cheeky sense of humor. The next day he handed me a little gift: a plastic keychain with a label for the driver's name, which he'd found at the general store in Millaa. In lieu of my name he'd written three words of advice: "THINK—DRIVE LEFT!"

In many ways Will was unique. I can honestly say, for example, that I've never met anyone since who carried a typewriter in his backpack. But Will, an aspiring author, did so, and clacked out long passages and letters to friends with impressive regularity. He possessed that sense of awe and magic that takes hold of some people, driving them beyond the normal limits of human endeavor. At times, manically energized by some recent discovery, mere words failed him. "The rainforest, Bill . . . It's just so, so . . . so aaaaaahhhhh!"

I nodded. I thought I knew what he meant.

I also came close to murdering Will a few times, and so did the other members of our crew. Although we really liked the guy, he never seemed to stop arguing. It didn't matter what the subject was—politics, biology, personal grooming habits—Will loved to take a contrary tack. If you thought the 49ers were going to win the Super Bowl this year, Will would give you five cogent reasons why they couldn't. Thinking about investing in the stock market? He'd be happy to talk you out of it, illustrating his arguments with frightening true stories of small investors who'd lost every penny in some market crash. He had a mind like a steel trap, and a mouth like Howard Cosell.

Will was the only person who ever argued with me about setting traps. We'd developed a ritual of walking every trapline at the beginning of each five-day session to inspect all the traps that had been laid out and baited. Setting traps is actually quite tricky. The trap needs to be stable and set properly. Because most mammals exhibit wall-hugging behavior—they like to run alongside logs and other objects—it's important to position the trap at the base of a tree or beneath a log or clump of vines. Ideally, there also should be some cover over the trap, like a shrub, so the animals don't feel exposed to owls, their main predators.

So I'd walk down each line and ask someone to shift this trap a few feet, or double-check that trap to make sure it worked properly. No one ever minded—until Will came along. Then even a simple request became a major drama. If I asked Will to move a trap under a nearby log, he'd launch into a long-winded explanation about why his trap placement was flawless as it was. He'd go so far as to crouch down on his hands and knees to illustrate just what the mammal would be seeing and thinking as it approached his trap. If I insisted, Will would get his back up and stubbornly refuse to budge. He was *sure* he was right and wasn't going to move the damned trap just because it was my project and I was telling him to do so.

Caught off guard by Will's defiance, on several occasions I found pleasant homicidal images flitting through my mind. What was I going to do with this guy?

I finally hit upon the answer: wrestling. It was the perfect solution. When I felt like killing Will I simply took him down and crunched him into a pretzel. Although Will considered himself a fair wrestler, not only was I bigger than him, but I'd wrestled in high school and college. This afforded us a number of opportunities to examine Will's attitude problem. I'd rack Will with a double arm-bar, then inquire as to whether he might reconsider moving the trap we'd just spent ten minutes arguing about. Or I'd torque him with the "guillotine"—my personal favorite—and ask if he wished to rethink all the insensitive things he'd said the previous evening about the San Francisco 49ers.

In wrestling, at least, Will lost his argumentative nature. He was too busy screaming and begging to be let go.

❦

We were soon joined by another volunteer, Jeff, a rotund, bearded Canadian, who brought our crew up to ten. Although we really didn't need an additional person, Jeff had his own jeep, which I coveted.

Like Will, Jeff turned out to be a character. Though lacking scientific training, he was the most enthusiastic advocate of evolution I've ever met.

To him, Stephen Jay Gould attained almost deity status. Jeff would propound at great length about evolutionary phenomena or bitterly castigate the fundamentalists who opposed teaching evolution in schools. He was fond of quoting Gould at length, having memorized his writings like some Confucian tome. He had a derisive attitude toward most people but liked me because I knew enough about evolution to be worth talking to.

For some reason, Bridgette began to seriously dislike Jeff. She described him as "creepy" and "weird." This was an overreaction, I felt, but nevertheless I wasn't pleased about this obvious discord among the troops. Will continued to drive us all crazy, as usual, but then we hit upon a solution that made everyone happy. Will and Jeff could work together—in isolation from the rest of us.

It was an ideal plan. Jeff had his own vehicle and we were just starting a new kind of trapping strategy, using extra cage traps I'd borrowed from the Tropical Forest Research Centre in Atherton. Instead of baiting these traps with our usual oats and vanilla, we would use beef or chicken. With these new traps we were hoping to catch predators—especially the spotted-tailed quoll, an almost mythical beast which had so far eluded us. A much-bigger relative of the ferocious antechinus, the quoll was the size of a big raccoon, all teeth and fury. It was the largest marsupial predator in the region—at least since the thylacine, or marsupial wolf, had been driven extinct.[1]

After a short training session, Will and Jeff headed off with the predator traps. The rest of us worked harmoniously doing our usual trapping and spotlighting. While I initially had reservations about turning Will and Jeff loose unsupervised, they did a fine job, and took great pride in their work. Jeff, in fact, liked to brag that they were "The Higher-Mammal Crew"—a reference to their working with glamorous predators—while the rest of us were merely "The Rat Patrol." That was fine with us. Will and Jeff were easy to take in measured doses—just not twenty-four hours a day.

It was now early October and the Austral summer was fast approaching. With the warmer weather we began to encounter more snakes, which were entering their breeding season. This was the time that most people got bit-

1. The thylacine's demise began with the arrival of dingoes, which were transported to Australia by Aborigines some 3,500 years ago and directly competed with thylacines for prey. By the time Europeans arrived in Australia in 1788, the thylacine had disappeared from the Australian mainland and New Guinea and survived only in Tasmania. The Tasmanian thylacine population succumbed to intense persecution from sheep ranchers and, quite possibly, to an exotic disease. The last known thylacine died in the Hobart, Tasmania, Zoo in 1936.

The Rat Patrol, ready for action. Pictured are Sally Beatty, Steve Comport, John Vollmar, myself, and Will Chaffee. PHOTO BY A BYSTANDER.

ten, when the snakes became aggressive and almost oblivious at times to people hiking in the forest. There had just been a horrible incident in Innisfail, a coastal city to the east of us. A toddler had been killed by a taipan while playing in his own backyard. He'd been bitten over a hundred times.

One morning we came almost face to face with a big red-bellied blacksnake—a good six feet long—which had reared up like a cobra upon hearing us approach. We paused and gave it plenty of room until it continued on its way, but it reminded us that we needed to stay alert.

Our next encounter was a little more scary. We were checking our traps one morning, numbed by the early hour. We were working in a hundred-acre fragment that, like many others, had been battered by the recent cyclone. Fallen trees, wait-a-whiles, and debris littered the ground. I reached down to pick up a cage trap that had been tripped, but couldn't see the trap well because it was obscured by debris. My hands were maybe six inches from the trap when a jet-black head shot out and struck at my hand, sending me flying backward. Everyone jumped as I hollered. A red-bellied blacksnake had gotten itself caught in the cage trap and was frantically trying to get out. It kept jabbing its head and neck out of the inch-square mesh but couldn't squeeze the rest of its body through the holes.

It took us quite a while to figure out how to let the snake go without getting bitten. We thought about killing it—it obviously lived right in the middle of our trapping grid—but decided that wasn't the right thing to do. We finally managed to release it with two long sticks, unlatching the trap door and holding it open long enough for the snake to spring to freedom and disappear into the undergrowth. We were all pretty jumpy after that—and completely wide-awake.

The worst snake encounter happened one night while I was out spotlighting. I was hiking along a trail that bisected a fifty-acre fragment. As was often the case, I was alone, and hadn't even brought Tully along. I'd just seen a female tree-kangaroo and her very cute joey foraging down on the ground, and was filled with warm fuzzy feelings about the rainforest and its amazing wildlife.

The trail I was hiking along was steep. I came to an eroding bank that dropped off for four feet. I grabbed hold of my backpack to keep the heavy battery from slamming into my back, then jumped off the bank, landing in soft dirt. Something squirmed and struck at my feet. I must have set some sort of world vertical-leap record—I'm not sure how high I went but I landed at least four feet away. I'd pivoted 180 degrees in mid-air and was now facing back toward the bank, shining my light where my feet had landed.

There in my spotlight, partly compressed into the dirt, were two extremely angry red-bellied blacksnakes. They'd been—in the professional lingo of biologists—doing it, when I'd had the extremely bad manners to fall out of the sky and land on them. They were now going completely crazy, thrashing and hissing and striking frenetically while they tried unsuccessfully to get away, their bodies firmly stuck together by the male's hemipenis.

I backed away quickly, my whole body shaking. A few moments later I stepped on a fallen branch and nearly leapt out of my skin again. *Now calm down*, I told myself, don't do a Joachim out here. I took deep breaths and eventually my heart stopped beating so wildly.

Finally I calmed down enough to resume spotlighting. In a way I felt rather sorry for the two blacksnakes. Here they are, enjoying a romantic evening—the moon is full, the night is warm—when suddenly this two-hundred-pound gorilla comes leaping out of the night and drowns them in the dirt. Talk about coitus interruptus.

something's in the bathtub

"When I'm finished with you, Laurance, you'll be reduced to a hollow, burnt-out shell of a human being. I'm done toying with you—tonight I'm going to shatter your fragile ego once and for all. After this you'll have to supplicate yourself before me and acknowledge that I am truly the King, the Emperor, the *God* of the Ping-Pong table."

Playing Ping-Pong with Will was always interesting. It entailed the kind of bluster and hype that one normally associates with Mohammed Ali before a big boxing match. Every match was a grudge match, every game an epic battle between the forces of good (him) and evil (me). It did lend a certain drama to the game. With Will, you always tried that little bit harder to win.

Of course, you had to. The bastard was good.

The only way to deal with Will was to go for the jugular. Before coming to Australia he'd applied to a number of prestigious Ivy League schools—Harvard, Yale, and Princeton among them. He'd arranged for them to send his mail here to Millaa Millaa, and each time a letter from one of these universities arrived there was a great drama as he opened it to see if he'd been accepted. So far, he'd had an unbroken string of rejections, and he wailed in disbelief with each new disappointment.

"Listen, Will, it's pretty funny to think you've ever shown *me* any mercy. I've only been taking it easy on you these past weeks because I know how downtrodden you've been about all these rejection letters. You might as well face it, mate—*you're just not Ivy League material.* If I were you I'd start applying to places like Chico State and Prairie Valley Community College. There, you might stand a chance of getting accepted."

These were fighting words. Whipped into a froth of aggression, we played as if it were the finals at Wimbledon, minus the decorum. Howling, screaming, diving for shots, our matches became so loud and dramatic that the others were forced to drop what they were doing to watch.

Neither of us ever really prevailed in these battles. We fought toe to toe, neck and neck, and ended up winning about half of the matches each.

And lest anyone feel concerned about young Will's education, I am happy to report that a few years later I received a long letter from him, elaborately typed as usual. After two years at Boston College he'd been accepted in the Department of English at Harvard. He was writing to tell me he'd just won Harvard's Literary Medal, awarded each year for the outstanding essay by an undergraduate.

Oh, and by the way, he said, guess what I wrote about? My time in Millaa Millaa. In fact, you, Bill, were a big star of my story.

Nice, I thought. Then the flattery faded to concern as I read the last line of his letter. "By the way, do you know any good defamation lawyers?"

Will and Jeff—the Higher-Mammal Crew—had been trapping for weeks but still hadn't encountered a spotted-tailed quoll. They had, however, caught several water rats with their meat-baited traps. The water rat is a bizarre animal, seemingly more otter than rodent. It has a torpedo-shaped body, luxuriously waterproof fur, webbed hind feet, and ears and nostrils with diaphragms that squeeze shut while it swims. It is also an aggressive carnivore that preys on crayfish, frogs, and almost anything else it can catch.

The water rat is yet another example of the unique Australian fauna. Like a giant floating island, the Australian continent slowly drifted in isolation for over forty million years. Many animals that are found elsewhere—dogs, cats, and monkeys, for example—never reached its shores. Instead, Australia became dominated by marsupials and by primitive, egg-laying monotremes—represented today by the echidna and duck-billed platypus, a kind of evolutionary link between the mammals and their reptilelike forebears. The biota is so unusual that it prompted a young Charles Darwin to muse that "two distinct Creators must have been at work"—one for Australia and one for the rest of the world.

Although completely isolated for much of its history, during the last fifteen million years the Australian continent has been slowly crashing into the Sunda Shelf, the vast tectonic plate that contains Southeast Asia. The island of New Guinea is a product of this momentous collision—rising out of the sea in a violent burst of volcanism and tectonic buckling.

As the distance between Asia and Australia gradually diminished, a few placental mammals—bats and rodents—began arriving in Australia. The bats flew across the sea while the rodents, presumably, inadvertently rafted on floating logs. These mammals had evolved the placenta, which is a profound reproductive innovation—the difference between a Lear jet and propeller-driven airplane. The placenta's function is to form a barrier between the female and her developing embryo. It allows food and oxygen to pass to the embryo but keeps the mother's blood and that of her offspring separate. This is a vital feature, for the embryo is genetically different from the mother—a mix of both maternal and paternal genes—and if their blood were to mix the mother's immune system would become activated and kill the embryo. For the embryo to develop to full term inside the mother's body, the placental barrier is essential.

Marsupials have only a rudimentary placenta, and for this reason the young cannot develop completely inside the female.[2] Just as the mother's immune system becomes activated, the young—little more than naked embryos—are born. They must crawl from the birth canal, up the mother's stomach, and into her pouch. There, they attach to a nipple and spend several months completing their development, warm and protected—but safely outside the mother's body.

When the rodents first arrived in Australia a few million years ago they found a number of vacant habitats, or ecological niches. One of these was water. Unlike placental mammals, which include aquatic species such as otters, sea lions, and whales, marsupials have never been very good at exploiting water. In the entire world, there is only a single semiaquatic marsupial, the water opossum of South America. The reason for this, I suspect, is that water would drown the young in the pouch.

The water rat took advantage of this glaring opportunity. It became a fearsome aquatic predator, superbly adapted for living in streams and rivers. Like the ungainly tree-kangaroos, however, the water rat faced a few constraints. One of these was its teeth. Rodents lack canine teeth, which other mammalian carnivores use to grasp and kill their prey. Instead they have a pair of chisel-like incisors on their upper and lower jaws, used for gnawing seeds and grasses. The water rat solved this problem by evolving razor-sharp incisors, the adjoining teeth growing so closely together that they essentially form a single, daggerlike fang. These fangs were renowned

2. This is not to imply that marsupial reproduction is inferior. In Australia's harsh deserts, for example, marsupials may have a reproductive advantage over placental mammals because they devote less energy to their offspring and can rapidly replace lost young.

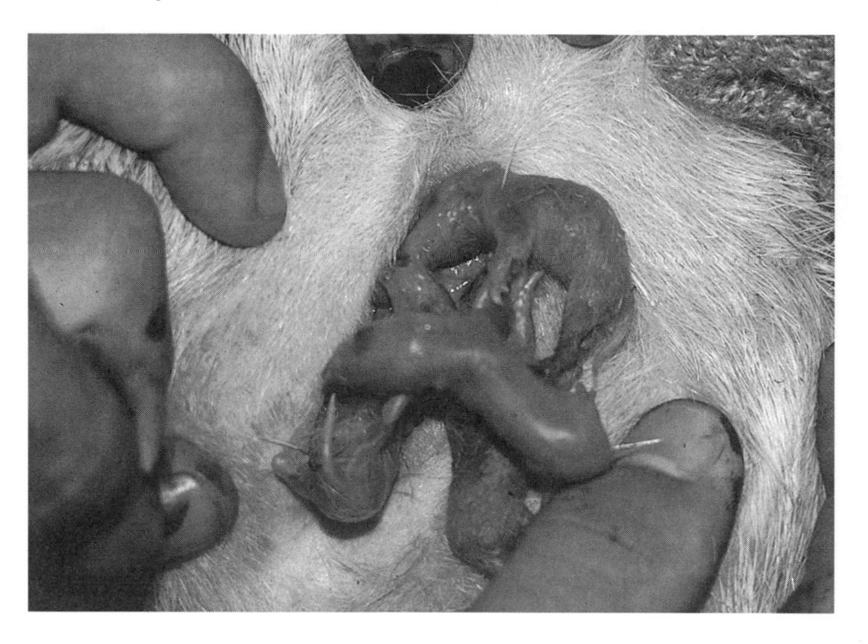

Pouch-young of a long-nosed bandicoot, showing the highly immature state at which marsupials are born. PHOTO BY AUTHOR.

among Australian biologists. Les Moore, perhaps the best field biologist in Queensland, once told me that the water rat was the only mammal he truly feared.

The first time Will and Jeff caught a water rat, they brought it home to show us. I walked into the house to find them looking enormously pleased with themselves. They refused to say anything, just pointed to the bathroom. There, swimming around in circles in the bathtub, was a huge water rat, even bigger than the giant white-tailed rats we'd been catching.

We all watched it for a while, admiring its sleek fur coat and remarkably otterlike behavior. It was surprisingly calm. Will told me they would let it go the next morning, and I reminded him not to put too much water in the tub, or it might be able to climb out. He assured me there was no way this could happen.

I was just about to retire for the night when Bridgette dashed in breathlessly to tell me the water rat had disappeared. Will and Jeff were desperately trying to find it.

We searched the entire house. It finally turned up in one of the bedrooms, hiding under a bed. It had backed itself into a corner and its formerly calm demeanor had vanished. I tentatively reached out my hand and it snapped viciously at my fingers, its growls unnervingly loud—more like an angry badger than any rodent I'd ever heard.

I did not have a good feeling about this. I'd never handled a water rat, and certainly not one in such a prime defensive position. Silently cursing Will and Jeff, I decided to don a pair of thick leather gloves I normally wore while clearing trails.

Taking a deep breath, I lashed out and grabbed. For a moment I had him, but he twisted free, his weasel-like body contorting unlike any rat I'd ever handled. His jaws opened and his long fangs sank into my hand, slicing through the glove like paper, his upper and lower fangs meeting in the center of my thumb.

I howled and retracted my bleeding hand, letting fly with a string of unprintable words. They say if you get thrown off a horse you should jump right back on, or you'll never get over it. That may be so, but there was no way in hell I was going to tangle with that water rat again. In my view, they rank right up there with death adders and giant centipedes.

Will and Jeff finally captured the rat by prodding it into a cage trap with a broom. My thumb was throbbing so badly I needed a couple of beers just to get to sleep. As I was turning in Will remarked that he and Jeff would be happy to bring home a few more water rats for me to handle—obviously I needed the practice.

Now and then life presents us with a few delicious ironies. Though I didn't know it at the time, the incident with the water rat was one of those moments. Will thought it hilarious that I'd been cowed by a water rat, but little did he know that his own rodent-Waterloo was soon to come.

The rainforest issues just kept getting hotter. Australia's prime minister, Bob Hawke, was being pressured to make rainforest conservation a key part of his bid for reelection, while Joh Bjelke-Petersen, Queensland's antienvironment premier, was threatening to run against him. Because of all the political overtones, the local media were keen for stories about the rainforest. I'd recently been interviewed by a regional magazine and the resulting article—"They Kill Rainforests Don't They?"—was written as a compelling plea to protect Queensland's remaining rainforests.

I was contacted soon afterward by a regional TV station that wanted to do a story on our research for their weekly news-and-variety show. We arranged a time for the film crew to visit us and promised to catch a few impressive animals for the cameras.

On the day of their visit we checked our traps as usual, but instead of releasing all the animals, kept two behind, a docile bush rat and a monstrous white-tailed rat, one of the biggest we'd encountered. We gave them food and water and covered the traps with leaves so they wouldn't get too warm.

The TV crew was supposed to arrive by noon or so, but they were running late. First they phoned to say they wouldn't be there until 4 P.M., then 6 P.M., and finally said they'd *definitely* arrive by 8. By the time they showed up it was pitch black. We grabbed our spotlights and with the TV crew in tow, headed out to the rainforest.

The filming started off well. We took them spotlighting and saw beautiful Herbert River ringtail possums, jet-black with white splotches. I told them of the intensive clearing and logging of forests in the region, and described the mammal species we'd found to be especially sensitive to habitat fragmentation. We then took them to our traps. In the eerie gleam of the spotlights, they filmed us weighing and tagging the bush rat, then releasing it to scamper off into the forest.

Finally it was time for the massive white-tailed rat. Having just spent twenty hours in a trap, the animal—an aggressive male—was not happy. He eyed us murderously and growled even more fiercely than usual.

The plan we'd devised was that I was going to hold the animal—I needed both hands for a beast this size—while Will tagged him. Unfortunately, two things went wrong at once. First, rather than hold the animal's neck, I decided to slide my hand down to its shoulders so its face would be clearly visible. Second, Will wasn't paying attention to what he was doing, but instead was gazing into the camera, no doubt fantasizing about his prospective career as the next David Attenborough.

Holding the tagging pliers, Will's hand moved toward the rat. It remained perfectly motionless until his hand was within range, then viciously chomped, biting straight through the skin and muscle at the base of Will's thumb.

Will screamed and swore as the pliers went flying.

I just stared in disbelief. Will was sucking on his hand and grimacing in pain. The cameras kept rolling but I knew it was to no avail. Despite the blood and drama, they'd never use the segment now, not with Will swearing at the top of his lungs. In a way I was glad they wouldn't; we'd probably look like hapless idiots.

The show was due to air that Friday. We all gathered at Des and Helen's house to watch it on their color TV. Everyone hushed when the story began. The narrator talked about forest fragmentation as the possums appeared on the screen. Then my interview, followed by the trapping scene:

the bush rat, then the white-tailed rat, then Will getting chomped and screaming at the top of his lungs "AAHHHHH—IT BIT ME! GODDAMMIT!!!"

We all roared. It was hilarious! We couldn't believe they'd used the segment in all its graphical glory, but they had. And in the end we didn't look like a band of idiots at all, thanks to the sympathetic way the piece had been narrated.

A week later one of the neighbor kids informed me that Will had been immortalized in the lore of the Atherton Tableland. At the local high school in Malanda, Will's spontaneous cry had become the cool thing to say. Whenever anybody got annoyed or hurt, they'd yell "AAHHHHH—IT BIT ME! GODDAMMIT!!!"

old harry

Studying the process of extinction is akin to trying to determine what caused a car wreck in which there were no survivors. Things have to be puzzled out afterward based on whatever evidence is at hand.

For us, part of the puzzle was trying to determine what the rainforests of the Atherton Tableland were like a century ago, before the arrival of European farmers and loggers. Which species existed then? Were the animal populations similar to those that occur today in the large forest tracts we studied—which survive only on steep mountainsides and poor soils? These questions are crucial, for they lie at the heart of our ability to understand the true effects of habitat fragmentation on animal communities. If we don't have a reasonable idea of what the forest was like *before* fragmentation, how can we possibly determine how fragmentation has changed it?

A case in point is the spotted-tailed quoll, the marsupial equivalent of a small bobcat, which so far had completely eluded our self-proclaimed Higher-Mammal Crew. Quolls are extremely rare on the Atherton Tableland these days. But were they more abundant before the rainforest was cleared?

I decided to approach Old Harry for help with this question. Old Harry was a placid, ancient Aboriginal man, rumored to be about ninety years old. I never knew him by any other name; that's what everyone called him. A member of the Waribarra tribe, he worked for decades scouting timber for the logging crews. The locals claimed that Old Harry knew more about the rainforests of Millaa Millaa than anyone alive. If he couldn't tell me what the rainforest was like before Europeans, no one could.

Old Harry was likely to be found in the pub, sitting silently, shoulders

The spotted-tailed quoll has virtually disappeared on the Atherton Tableland. Forest fragmentation and toxic cane toads seem to have caused its decline. PHOTO BY STEVE WILLIAMS.

hunched, nursing a beer. He was small-boned and thin, as is typical of the rainforest Aborigines, and his densely wrinkled, ebony-black skin contrasted starkly with his thin white hair.

I approached Harry in the pub one day and asked for a few words with him. He wasn't known to be much of a talker, but he listened thoughtfully to a description of our work in the area. He then explained quietly that the rainforests had indeed been profoundly altered. "Cassowary, red owl, big python. Dem ain't here no more. Dem used ta be here, but dem ain't here no more."

His description coincided to a remarkable degree with what our research was revealing. The huge, flightless cassowary had virtually disappeared from the tableland's fragmented forests. The formidable red owls—called rufous owls by birdwatchers—also seemed to have vanished. And large carpet and amesthytine pythons, some over twenty feet in length, had become extremely rare, although much smaller pythons were still present.

Old Harry also confirmed my suspicions about the quoll. It had been quite common prior to World War II, frequently raiding henhouses on farms. "Dem quolls like ta eat chooks," Harry told me—"chooks" being Aussie slang for chickens. But the quoll was no longer.

Old Harry was one of the last of the Waribarras, and so frail it appeared he wouldn't live much longer. He died only a few months later, and the locals erected a small monument in his honor near the Millaa Millaa waterfall. It was a delight to have shared a conversation with him, to have glimpsed a view of the rainforest past. And it turns out that Old Harry was well known among Australian anthropologists. His full name was Harry Digala, and when he died, Millaa Millaa's unique dialect of Dyirbal, the native tongue of the rainforest Aborigines, died with him.[3]

In the ancient tongue of Old Harry's tribe, "Millaa" meant water or rain, and when the Aborigines wanted to say "a lot" of something, they just said it twice. Hence, Millaa Millaa is the land of rain and water, an unquestionably apt description of the dripping forests of the southern Atherton Tableland. Like the rainforests that once cloaked the tableland, the Aboriginal tribes that populated the region have collapsed, just as they have throughout most of Australia.

For our research team, the depressing implications of massive forest destruction were reinforced daily. Even John, my happy-go-lucky assistant, was occasionally affected, and one day he shocked us all with a burst of strident idealism.

We'd had a hard time finding suitable study areas in the continuous forests near Millaa Millaa. Most of the unfragmented forests clung to steep mountainsides or overlay poor soils, factors that could bias our comparisons with the forest fragments, which mainly occurred in flatter areas on fertile basaltic soil.

We'd managed to find a nice patch of red basaltic soil on Mt. Fisher, but it abutted a tract of private rainforest—owned, in fact, by the same old logger who'd cussed us out so vigorously when we had inadvertently wandered onto his land. By now, late November, we'd been working at this site for over six months. As is probably typical of biologists, we'd begun to feel rather proprietary about our study area, having gotten to know it and its wild denizens so intimately.

In the midst of trapping one day, we were surprised by the sounds of crashing trees and the heavy growl of an approaching bulldozer. Within

3. This conclusion, and my interpretation of the Waribarra name for Old Harry's tribe, is based on the work of the cultural anthropologist R. M. W. Dixon (for example, see Dixon, "A changing language situation: The decline of Dyirbal, 1963–1989," *Language in Society* 20 [1991]:183–200).

minutes the old logger appeared on a large yellow bulldozer, happily toppling trees and clearing a track barely fifty yards from us, just on the other side of a creek that defined his property line. He waved cheerfully—we'd long ago established a casual rapport.

We stared at him, transfixed and horrified. It was one thing to talk about rainforest destruction, but something else to watch dozens of stately trees come crashing down before your eyes. John was swearing quietly, but we were stunned when he suddenly bellowed and took off straight for the bulldozer.

We bolted after him, screaming. He was yelling furiously as he ran, and I managed to draw close and grab a handful of shirt just as he started up the far streambank. He whirled around, tears in his eyes.

"Goddamit, he can't *do* that!" he screamed. "He'll screw up our whole study!" But of course it was more than that. It was the thought of someone bulldozing his way into our pristine little valley on the edge of Mt. Fisher. Right in front of our eyes! Though we all felt as bad as John did, there wasn't a damned thing we could do about it.

Fortunately, the logger hadn't seen John's quixotic charge. And I wasn't the least bit disappointed in John. If you dug deep enough, there was a fair-dinkum conservationist in there, and that always impresses me.

Some weeks later John nearly took his revenge. We were spotlighting on Mt. Fisher and came upon a shed where the logger parked his bulldozer and his skidder, a big rubber-tired vehicle used for dragging logs. For a moment we paused, and the same tempting thought must've flashed through both our minds. John said with a wicked smile, "Ever read *The Monkey Wrench Gang*?"

The weather grew hotter and muggier as December approached, and after months of intense fieldwork I felt like I was reaching my physical limit. One day I thought I'd gone too far. As usual, we were working hurriedly to finish trapping, so we could race home, peel off our muddy clothes, and collapse for a few hours. I was lugging a load of cage traps up a hill when a sharp pain shot across my chest and down my right arm. My arm went numb. I dropped the traps and rubbed my throbbing chest and shoulder. The chest pain was getting worse—like an intense cramp.

The rest of the gang stopped when they saw the alarmed look on my face. "You all right?" John asked, puzzled. These were the classic symptoms of a heart attack, I realized, but I refused to believe it. Sure, I smoked cigarettes and probably worked too hard, but I was fit and strong and in my late twenties. A heart attack?

The pain forced me to sit down. Though sweating heavily, I felt strangely chilled. I kept kneading my chest. Someone handed me some water, and after several gulps I felt for my pulse. It was very rapid. Was that good or bad? I tried taking deep breaths—calm down, calm down. But the pain in my chest was growing stronger.

Then I felt a bump. Protruding from my chest, the size of a small grape, was a blood-engorged tick that had burrowed its head into my skin. A paralysis tick! I ripped it off, and felt a huge surge of relief. If left untreated, paralysis ticks can be fatal, but I didn't think it'd been there long enough to cause permanent damage. I wasn't in the throes of cardiac arrest!

My chest and shoulder were sore for days; it's remarkable the wallop a little tick can pack. Indeed, paralysis ticks have been known to kill animals as large as horses and cattle. Our only prior encounter with a paralysis tick was when we found a white-tailed rat flopping around spastically in one of our traps, as if in the throes of an epileptic fit. We managed to locate and remove the tick that would have soon killed it.

We learned later that paralysis ticks plague many rainforest animals. Ticks that cause paralysis occur throughout the world, though only a limited number of species are known to do so. The paralysis is apparently inadvertent; the toxin injected into the host is part of the tick's salivary fluid, which improves the flow of blood to the feeding tick. The toxin disrupts the junction between nerves and muscles, so that the victim goes numb and loses coordination. If the tick is found and removed in time, the toxin's effects are reversible. But if the tick remains, the paralysis eventually spreads to the breathing center of the brain, causing death by suffocation.

As part of their efforts to pressure the federal government into designating the rainforests of north Queensland a World Heritage Area—effectively closing down the logging industry—the conservationists in north Queensland decided to place a very public phone call to Prime Minister Hawke. All of the television stations and newspapers would be invited, to help draw attention to the strong regional commitment to rainforest conservation.

Because a recent lecture I'd given for the Rainforest Conservation Society had been well received, I was asked to make the phone call. On the appointed day I drove down to Cairns, dressed in my best (only) sports jacket. I even wore a tie—almost unheard of for me. The idea of talking to the prime minister made me nervous, but I'd prepared a little speech: a list of reasons why I thought it was vital to protect the region's rainforests.

The circuslike atmosphere I encountered on my arrival caught me completely off guard. I'd expected the whole thing to be sober and serious, but the telephone where I was to place my call had been surrounded with garish posters bearing strident rainforest slogans and myriad flowers and vines. Some bloke dressed in a platypus suit insisted on sitting right next to me, his arm symbolically around my shoulder as though we were best mates. One of the leaders of the local conservation movement was hyping the media like a circus barker.

As it turned out, I didn't get to talk to the prime minister, just one of his staff. Apparently, this had all been arranged in advance without my knowledge. The staffer instructed me to pretend that *he* was the prime minister, and he'd pass on our very, very important message—which the prime minister was just dying to hear—straight away, personally.

I chatted away while the cameras rolled and the reporters scribbled—I tried to look like I was having a very thoughtful and important conversation with the democratically elected leader of Australia. I began with a half dozen reasons why rainforest logging should be stopped, ranging from the unique biological significance of the region to the fact that, because of past overcutting, the industry was gradually collapsing anyway. All the while I was cooking beneath the hot TV lights in my jacket and tie and my chest was still hurting like hell from the damned paralysis tick and this big ape in a platypus suit was practically sitting in my lap. The torture session finally ended. Just as I was preparing to flee, another TV crew arrived and asked if we could repeat the whole performance. By now I was quite accomplished at having imitation conversations with imaginary prime ministers so I didn't even bother to phone the staffer back; I just faked the whole thing. Everyone who caught the broadcast said I looked like I really was having this totally pithy and absorbing conversation with ol' Hawkie. I was sorry to have to tell them the truth.

December arrived and our plans for Christmas began to coalesce. Because we were all worn out, we decided to take a real break: two weeks in Sydney. As the holiday season drew near, our crew had dwindled to four—John; two British women, Sue and Gail; and myself. Dale and Cathy, happily reunited, now lived in a flat near Bondi Beach, Sydney's famous surfing area. We were invited to come spend Christmas with them, an offer too good to refuse.

We finished trapping and stowed away our equipment. The neighbor kids were happy to look after Tully; they practically did that anyway. We were loose and happy, the prospect of a few weeks away from mud and leeches just the tonic for our ills.

I made a final run to the post office, leaving the gang to finish loading the truck for the two-thousand-mile drive to Sydney. We would travel straight through, alternating drivers, stopping only for meals and a little birdwatching along the way.

At the post office a letter was waiting from Beth. My throat instantly tightened. What could *she* want? Some deeply buried part of me hoped she wanted me back. But it'd been almost a year since we'd split; why now?

In the pretty little park on Millaa's Main Street, I took a deep breath and ripped open the letter. She was fine, her folks were fine, she was starting graduate school . . . she'd met someone and it had gotten serious. They were going to get married. She wanted me to know.

When I finally got back to the house the gang was raring to leave. I asked John to go ahead and drive; I was tired. I climbed into the back and sacked out on a mattress, where I slept most of the way to Sydney.

Christmas in Sydney was a blur. We carried a Christmas tree down to Bondi Beach and slowly drank the day away. There were lots of zany antics and we tried not to stare too obviously at the beautiful women toplessly sunning themselves, but deep down I think most of us were a little melancholy. We were all away from family, and it's hard to get in the holiday spirit without snow. To top it off, our attempts to swim were thwarted by the territorial teenage surfers.

Most of our gang had made a one-way trip to Sydney; they wouldn't be returning to Queensland. John would be coming back eventually but was short of cash; he would remain in Sydney and try to find work as a waiter. We had a New Year's Eve party at Dale and Cathy's flat, saying our goodbyes the next day. I departed alone for north Queensland, the width of a continent away. On the way back I planned to stop in Brisbane to visit the queen of Australian conservationists.

To friends and foes alike, Aila Keto is a formidable woman. A former biochemist at the University of Queensland, Aila quit her academic job to become a full-time conservationist. Her husband, Keith, also an academic, resigned as well, and together they formed the Rainforest Conservation

Society of Queensland, based out of their Brisbane home. With Aila as its figurehead, the RCSQ did more than any other Australian organization to promote the conservation of north Queensland's rainforests.

A woman of remarkable tenacity, Aila has cowed the toughest politicians. Though she is invariably polite and soft-spoken, few can withstand her steely resolve and withering barrage of hard, cold facts. Aila combines the best features of a zealot (an unshakable belief that she is right) and a first-class academic (an understanding that knowledge is power). Aila wins because she's smarter, more committed, and more disciplined than her opponents.

I'd contacted Aila before leaving for Sydney and was immediately invited to spend a few days with her and Keith in Brisbane. They were gracious hosts, and we shared the kind of trust that comes from knowing we were staunch allies at heart. Aila arranged a radio interview for me in Brisbane and I put forward the case for conserving north Queensland's rainforests. I tried to emphasize their biological significance not just to Australia but to the entire world.

the wet season

By early January, the wet season had Millaa Millaa locked in its soggy embrace. The temperature plummeted about thirty degrees Fahrenheit and the chill from the constant driving mist seemed to penetrate one's bones. The sun disappeared behind gray, sullen clouds for weeks on end. It may sound odd to say a person could suffer from hypothermia in the tropics, but it's entirely possible in the wet season, especially at higher elevations. The postmistress complained incessantly about the cold and mud, swearing, as she'd done for years, that this was absolutely the last time she'd ever spend a wet season in Millaa Millaa.

For months moisture pervaded everything. I was dismayed to open my closet one day and find that all four of the decent shirts I'd brought with me were encrusted in black mold. Leather shoes had also fallen victim. Rot and mildew seemed to grow everywhere—on the walls, our clothes, our bodies. Two leech bites on my leg became infected with jungle rot—a virulent fungus—that seemed impervious to medication. The fungus formed deep, puss-filled cavities that lasted for weeks.

Almost as bad was crotch rot—a scarlet rash that colonizes the groin and itches maddeningly. Throughout the wet season our gang was readily distinguished from normal folks by a propensity to scratch our private parts at any moment. We itched and rubbed ourselves raw, so much that we became almost unaware that we were doing it. Once I was chagrined to find I'd been mindlessly scratching my groin while chatting with Helen in the takeaway.

We attacked the crotch rot with antibiotic and antifungal creams, but none provided more than momentary relief. We were slogging around eight

TOP An Australian volunteer, Paul Downs, gazes at one of the Atherton Tableland's many waterfalls. PHOTO BY CHRISTINE BUEHLER.

BOTTOM During the wet season, small creeks turned into wild torrents. PHOTO BY AUTHOR.

or ten hours a day in soaking wet clothes, in jungle conditions—fungus heaven.

Finally one of the local farmers told us his antifungal secret—copious amounts of baby powder. Each morning we'd queue up at the bathroom to douse ourselves in great clouds of delicately scented powder, before heading to the forest. To our great relief, the fungus that had been eating us alive was finally brought under control.

Our highest priority upon returning to the north was to find new volunteers. John returned sooner than planned from Sydney, his efforts to work there unsuccessful because he lacked the proper visa. We phoned the Cairns hostels, asking them to keep their eyes open for potential volunteers, but the pickings were sparse; tourism in north Queensland plummets during the wet season.

We managed to pick up two good helpers—Jeremy, a classy English gentleman in the finest sense; and Ingela, a placid Swede. A third volunteer, a jolly Irishman named Patrick, went to great lengths to rearrange his plane tickets so he could spend a month with us. But after a single week of mud, rats, and rain, he decided he'd made a horrible mistake, and left. I was annoyed with him, both for failing to stand by our bargain and because volunteers had been so hard to find. But on his last night he handed me a cassette labeled "Ode to Bill." A guitarist, he'd written and sung me an Irish ballad, the essence being that while he greatly admired the work we were doing, it just wasn't for him. The song was funny and poignant, and I found I just couldn't hold anything against him.

We struggled through January with only a skeleton crew. Then we received a call from an Australian, Steve Comport, who'd heard about our project and was determined to work with us. He arrived in Millaa in his white "combi" camping van, in which he'd been touring Australia for the past several months.

At first we weren't quite sure how to take Steve. He was certainly a bundle of enthusiasm. During our initial meeting he talked nonstop, regaling us with tales of parachuting—he was a sky diver with five hundred jumps to his credit—and with horrifying stories of near-death experiences.

At the tender age of twenty-three, Steve had suffered a sort of premature midlife crisis. He'd left high school to become a machinist, then spent seven years in jobs that left him increasingly bored and dissatisfied. Always intrigued by animals and biology, one day he simply quit his job, loaded up his van, and took off on a great sweeping tour of Australia, determined to find a research project on animals. His prayers were apparently answered by our sign in a Cairns hostel. We probably couldn't have kept him away with a Sherman tank.

Almost overnight, Steve made himself indispensable. He was a natural organizer and planner, and a perfectionist—traits that distinguished him favorably from John and his sometimes laissez-faire way of doing things. Steve imparted an almost military sense of efficiency to our operation. This sudden new discipline extended even to Tully, who'd become accustomed to lolling about the house and begging for table scraps. Steve put a quick stop to all that, and his irritated yells of "Out! Out!" became a regular household feature.

I was in heaven. Steve was just what I'd been needing—someone to tackle all the day-to-day operational details with enthusiasm. This would free me up to focus on the science and, increasingly, on a need to find additional funds to keep the project going. Within a month I'd designated Steve my official "Logistics Coordinator," a post he assumed with great pride.

The wet season brought with it a sudden profusion of cane toads. As an island continent, Australia's ecology has been drastically altered by introduced species—foxes, feral cats, and rabbits being three of the worst. As a result of these foreign species, Australian mammals have suffered a higher rate of extinction over the last two centuries than those of any other continent. Most of the species that disappeared lived in arid regions and were small—less than a few kilograms—perfect-sized prey for foxes and feral cats.

The cane toad is another sad chapter in Australia's ecology. A native of the New World tropics (today, at our home in the Brazilian Amazon, cane toads breed in our backyard), the cane toad was released into north Queensland in a misguided effort to control the greyback beetle, an exotic insect whose larvae cause serious damage to sugarcane. The cane toad was utterly ineffective in controlling the beetle, which spends most of its life above ground—well out of reach of the squat, ground-dwelling toad. Even worse, the toad—which has to rank among the ugliest of all God's creatures—bred prolifically and began spreading across the Australian continent at a rate of two miles a month. At the front of the invasion wave, the cane toad achieved extraordinary densities—sometimes hundreds of animals per acre.

The cane toad has had a dramatic impact on the native fauna. In such great numbers, it greedily consumes all potential prey in its path. The toad prefers insects and other invertebrates, but will eat virtually anything—frogs, small marsupials, rodents. It even likes dog food, and has been known

A noxious invader from South and Central America, the cane toad is deadly to native predators and is rapidly spreading across northern Australia. PHOTO BY AUTHOR.

to queue up patiently at Fido's bowl every evening to be fed. In addition, the toad is deadly poisonous. It sports two large bumps, called paratoid glands, on the back of its head. When the animal is alarmed, these glands secrete a milky ooze that contains a cocktail of powerful cardiac poisons.

With its potent weapons, the cane toad doesn't flee potential predators; it just sits there, daring any animal to tangle with it. Unfortunately, the Australian fauna evolved without toads, and thus lack any natural defenses against them. It is impossible to calculate the number of Australian predators that have died attempting to eat cane toads, but anecdotal evidence suggests it must be massive. For example, two different farmers on the tableland told me they'd found dead pythons with cane toads in their gullets. An Australian biologist, Michael Archer, found his beloved pet quoll trying to eat a cane toad in his backyard. The quoll went into spasms and, in just minutes, died in his arms. Populations of monitors—big lizards related to the Komodo Dragon—have been shown to decline sharply as the invading wave of toads passes over them.

Today, the cane toad has become a fixture of Australian culture. Some (clearly misguided) souls actually *like* the toads, allowing their children to keep them as pets. In the hilarious Australian documentary *The Cane Toad: An Un-natural History*, a young girl dresses up her pet toad and pushes it around in a baby carriage. Another vignette involves a drugged-out hippie who likes to smoke the dried, hallucinogenic skin of cane toads—the wastoid spouting Carlos Castaneda-like gibberish about "becoming one with the cane toad." My personal favorite is the guy who bought special wide tires for his van so he could efficiently smash toads while racing down the road.

On the Atherton Tableland, the cane toad may well have contributed to the demise of native predators such as the spotted-tailed quoll. The toad is most abundant in open areas like pastures, but it also invades rainforest along streams and, especially, along logging roads. On one wet night we counted fifty-two toads along a one-kilometer (0.6-mile) stretch of logging road. Sometime afterward I spoke to a forester who confidently assured me that logging had only a negligible impact on the rainforest. I asked him if he was aware that the hundreds of miles of logging roads in the region apparently acted as avenues for cane toad invasions. The thought had never occurred to him.

In January an American organization called the School for Field Studies established an outpost in Millaa Millaa. The new center was intended to provide semester-long field courses in rainforest ecology to American undergraduate students. SFS mainly accommodated students from well-to-do families, given that the tuition fees for one semester (not including airfares) were around ten thousand dollars.

SFS rented two houses a few miles from Millaa, and, under the directorship of a nice chap named Erv Petersen, began teaching their fifteen students about rainforests. We became quite chummy with the SFS mob, both because we were all Yanks living in Millaa Millaa and because some of my male volunteers became very friendly with some of Erv's female students. Erv took a dim view of this sort of fraternization, but, the forces of nature being what they are, had only limited success in heading off a couple of budding romances.

The SFS students were all nice kids, though initially a bit timid in the rainforest. On one of their first sojourns they joined us while we were trapping in a small rainforest fragment. It was raining thunderously, and most of the students looked miserable. By this time we'd become quite accustomed to working under such conditions, soaked to the bone. We couldn't help noticing that several students were decked out in the latest L. L. Bean fashions, with accessories like web belts, canteens, and hunting knives, which must have cost them hundreds of dollars. After a few minutes they dashed back to their van and headed home. Within a month, however, the students had become better acclimated and were tromping around the rainforest with confidence.

Like us, SFS had a big impact on the social dynamics of Millaa Millaa. Erv brought the students into the Millaa pub on Friday nights, and many of the locals were taken aback by this unprecedented Yankee invasion. Even during World War II, when thousands of Americans were stationed in north Queensland, few ever made their way up to Millaa Millaa.

The SFS students became friendly with several of the local residents. One of their most memorable friendships was with a diminutive dairy farmer by the name of Pat Reynolds. A happy-go-lucky bachelor, Pat packed more personality into his small frame than virtually anyone I've ever met. He was a raving extrovert, with strong views on most subjects, and woe be to those who tried to dispute him. At a barbecue I watched Pat debate the entire group of students about rainforest conservation, and he easily held his own.

Pat was also a notorious maker of practical jokes. The night before April Fools' Day a few years back, he'd snuck over to a friend's dairy farm and welded all his gates shut. The cows queued up at dawn, anxious to be fed and milked, but the poor farmer couldn't get a single gate open. Finally, with the cows about to go berserk, he panicked and smashed open his gates with a sledgehammer.

Foolishly, the SFS students got into a practical-joke contest with Pat. The way I heard it, the students started the battle by burning bags of cow dung on Pat's porch. This was a serious miscalculation. Pat phoned the students a few days later and asked them to stand outside their house and wave to a friend of his, who was spraying a nearby field in his biplane crop duster. The students went out, but the crop duster flew over at tree height and scored a direct hit on them with a load of orange dye.

But Pat wasn't finished. A few nights later, the students were woken in the middle of the night by screams of "Fire! Fire!" As they drowsily looked out their windows, Pat lit a flammable solution he'd poured all over the lawns and shrubbery. As flames surrounded the house the students panicked and fled—only to collapse in great piles as they hit the front porch, which Pat had doused in motor oil. This memorable event brought an end to the practical-joke spree, as the next day Erv told Pat in no uncertain terms that that had better be the end of it. As expected, this epic contest became part of the permanent lore of Millaa Millaa. Final score: Pat Reynolds 2, SFS 1.

No accounting of practical jokes Down Under would be complete without the story of "Utu" Bauer. This tale takes place not in Australia but in New Zealand, where the residents—perhaps also possessed by some devilish quality affecting those of the Antipodes—take an equally perverse delight in ornery humor.

For me, the story of Utu came to light in early 1985 as I chatted with my friend Aaron Bauer, a fellow grad student at Berkeley. Aaron was a mellow, likeable fellow, but one morning I heard him swearing bitterly as he checked his university mailbox. Curious, I glanced over his shoulder and saw a letter prominently addressed to "Utu Bauer." Though he initially didn't care to talk about it, after some pestering he finally consented to tell me how he'd earned that nickname.

Like me, Aaron was doing his fieldwork in the Southern Hemisphere. His project involved an elaborate study of a particular group of lizards—geckos. Aaron was trying to piece together the evolutionary history and biogeography of every single species of gecko in the South Pacific. To collect his specimens he had gone to extraordinary lengths, scouring the region's many islands for unique species, wading through murky swamps at night with a headlamp, even trying to persuade superstitious South Pacific islanders—who steadfastly believe that gecko bites are deadly—to climb palm trees to search for rare species.

Aaron's journeys had finally taken him to New Zealand, purportedly home to the most mysterious of all geckos—a giant species that was known only from Maori legends and folktales. A few months earlier, Aaron had been stunned to discover a specimen of the mythical beast (which had been wrongly identified) in the dusty recesses of a French museum. He and a colleague named it *Hoplodactylus delcourti*—and at nearly two and a half feet in length, it was by far the biggest gecko in the world.

Bursting with excitement, Aaron flew to New Zealand in the desperate hope of finding a relict population of the giant gecko. Although it hadn't been seen for over a century, Aaron was an optimist. He direly wanted a blood sample from the gecko for his DNA studies, and he was determined to visit any surviving areas of native forest from which the species had originally been known. Perhaps, he thought, he might just get lucky—and make scientific history while he was at it.

Aaron stayed with a friend of his, a fellow herpetologist in Nelson, on the South Island of New Zealand. Counting on the help of local residents, Aaron spoke to naturalists' groups and the media, then took several sojourns into the field. In New Zealand, much interest was raised in the search for the giant gecko—it even made the front pages of some newspapers.

There was no response initially, but when Aaron returned to New Zealand a few months later, his friend excitedly showed him a letter sent to him by an elderly farmer. The letter had been posted from the northern tip of the North Island, in a remote area that still retained some of its ancient kauri-pine forests. The farmer said he had been seeing giant geckos for years—and he described *Hoplodactylus delcourti* to a T.

Aaron was in agony. He was scheduled to depart immediately for New Caledonia, but here was evidence of the scientific discovery of a lifetime! He wanted to drop everything and head straight for the North Island, but finally decided to continue his travels when his friend promised to follow up on the tantalizing lead immediately.

Unable to contain himself, Aaron changed his plans so he could return to New Zealand a short time later, and he arrived to find his friend triumphant. His friend had immediately contacted the old farmer, who'd sent them a black-and-white photo that was taken years earlier. Aaron took one look and his eyes nearly popped out. There was the farmer—holding a gecko that was five feet long!

This went beyond Aaron's wildest dreams. Not only was the giant gecko apparently not extinct, but the massive specimen he'd discovered in France was just a baby! Marching bands played triumphantly and fireworks exploded as Aaron realized he'd just found his ticket to scientific immortality. For a rock musician, true fame means seeing yourself on the cover of *Rolling Stone* magazine; for a politician, it's the cover of *Time*. But for a scientist, the absolute pinnacle of achievement is making the cover of *Nature* or *Science*—the world's foremost scientific journals. Aaron was already imagining his cover photo: him smiling modestly as he held up a gecko the size of an alligator. His name would be on the lips of every biologist in the world!

But then Aaron noticed his friend's face was turning red. Suddenly his friend exploded in maniacal laughter. The penny dropped: Aaron's friend had engineered the whole damned stunt. It turns out he'd teamed up with friends to fake the elaborate letters and the remarkably realistic trick photograph—which was actually a blow-up of a much-smaller gecko species. He'd even sent the letters to a mate on the North Island to be postmarked and mailed back to him. Reeling as his visions of glory collapsed, Aaron managed to sputter just one word in reply: "Utu"—the Maori word for revenge.

But this tale is not quite finished. Aaron reports today that his Kiwi friends continue to rub salt in his wounds on every possible occasion. They even went so far as to change his name on a scientific paper he recently wrote for the *New Zealand Journal of Zoology*—to "Aaron Utu Bauer." Though it has been nearly fifteen years since he was so disastrously hoodwinked, he swears that one day he will have his revenge.

jenny

While our research was going fine, my personal life felt like a shambles. Romances seemed to be springing up all over, but none involved me. Though surrounded by a mob I felt lonely; it had been a year since I'd had a girlfriend. My loneliness was only worsened by our travelers, who came and left with depressing regularity. I'd just start forging a friendship with someone and it'd be time for them to leave.

One evening Des and Helen, the owners of the Millaa takeaway, invited me over for a barbecue. Helen and I were quite chummy, and upon arrival I quickly realized she had set me up on a blind date. I was paired off with a pretty young woman named Jenny Taylor, who had a six-month-old baby girl. Jenny had become pregnant during a five-year relationship which she'd ended just before the baby was born.

Jenny had a funny, cynical sense of humor, and I was drawn to her. Originally from Millaa, she'd spent a few years living on the Gold Coast down south, and was quite sophisticated. We laughed a lot and she teased me mercilessly for inquiring what her "baby boy's" name was—despite the fact that the little girl, Kasie, was decked out in pink from head to toe.

The next night I invited Jenny over for dinner, having emptied the house by giving Steve and the gang some money to visit a pizza parlor in Malanda. Within two weeks, Jenny and I were spending most of our free time together. She had her own house on the other side of Millaa, and I often stayed there in the evenings. Naturally we were the subject of considerable gossip, which bothered neither of us in the least.

Jenny's mother, Shirley, was a wonderful woman, a true pillar of the community. Whenever anyone was in need, Shirley was the first to arrive, spending her weekends caring for sick friends, babysitting, or helping neighbors to remodel their house. We became good friends, and I was delighted to learn that Shirley was sympathetic to rainforest conservation. It had always bothered her that so much of the region's rainforests had been devastated.

Like many of the locals, Jenny had grown up taking the rainforest for granted. The pioneering spirit that had conquered the tableland's forests was very much alive in the region's rural towns. Many of the original settlers, themselves only recent immigrants from Britain or Europe, had died as they fought to beat back the rainforest and the Aborigines to establish their farms. Others succumbed to diseases like leptospirosis or to deadly snakebites. To the settlers the entire landscape was alien, unlike the tidy villages and neat, hedgerowed paddocks of Europe. So they dramatically transformed it, making it more tame, more familiar, less threatening. The rainforest was the foe—"scrub," they called it. Nearly a century later these attitudes still persisted, despite the fact that the rainforest had long since been vanquished.

I began taking Jenny out spotlighting, while her younger sister looked after Kasie. A gradual change came over Jenny, which seemed especially evident one moonlit night as we watched a mother tree-kangaroo groom her darling young joey, which she held protectively in her arms. Something about the incident touched her deeply. As a young mother herself, I think she suddenly realized that the rainforest wasn't just useless scrub.

Although there was considerable tension surrounding rainforest issues, we'd been making a genuine effort to be friendly with the locals, with generally promising results. No one had ever refused us permission to work on their land, and I was even invited to give a talk before the Millaa Millaa Rotary Club. I didn't know much about the Rotarians other than that they were mainly farmers and businessmen and did good deeds like raising money to provide scholarships for local students.

Jenny and I were warmly received, and the Rotarians and their wives listened attentively while I described my research in a lighthearted way, telling anecdotes and showing slides of rainforest animals. I toned down my usually forceful conservation message, avoiding the topic of logging entirely but emphasizing the vital need to protect the remaining fragments of rainforest on the tableland. A strident talk would have alienated a group like this, and polarizing people is usually counterproductive. The evening was fun, and we seemed to have made headway with some of the influential locals.

Steve approached me a few days after my talk to tell me that a woman named Christine had called. She seemed to know me. She asked if I still needed help with fieldwork, and when Steve had said yes, she hung up.

Could it be? I asked if she had an accent and Steve said yes, definitely, German or something. I was ecstatic—I kept saying "Great!" over and over and telling everyone that if this was who I thought it was, she'd be a wonderful, invaluable person to have around.

Christine showed up the next day, backpack in hand. She'd worked several months in New Zealand and now had another six-month Australian visa. She'd been missing the rainforest. Could I use some help?

I'd hyped Christine so much that our gang was initially shy around her, almost suspicious. But within an hour she had them all won over; it was funny to watch. With her bubbly laugh and utter lack of pretentiousness, she was impossibly likeable.

Steve and Chris soon became good friends and with the two of them, I had a terrific combination. Travelers came and went, but they stayed for months on end and ensured that the project ran beautifully. John had left in January to search for work on the tableland, and we now tried to recruit older, less party-oriented travelers. We became practiced at interviewing prospective volunteers over the phone to determine if they had a sufficiently mature attitude.

Christine was an incurable animal-lover, and our house soon resembled a menagerie for hurt and orphaned animals. At any point we might be looking after a baby brushtail possum who's mom had been killed by a dog, a barn owl who'd been hit by a car, and a Boyd's forest dragon that had turned up disoriented in somebody's dairy shed. Tully proved to be quite tolerant of all these newcomers once he realized it was expected of him.

Because little was known about the diets of the rainforest rodents we were trapping, we decided to conduct some experiments to learn more about their food preferences. We built a half dozen plywood enclosures in a shed in our backyard, and into each we put a bush rat, Cape York rat, or fawn-footed melomys. Each animal was offered a smorgasbord of foods, ranging from

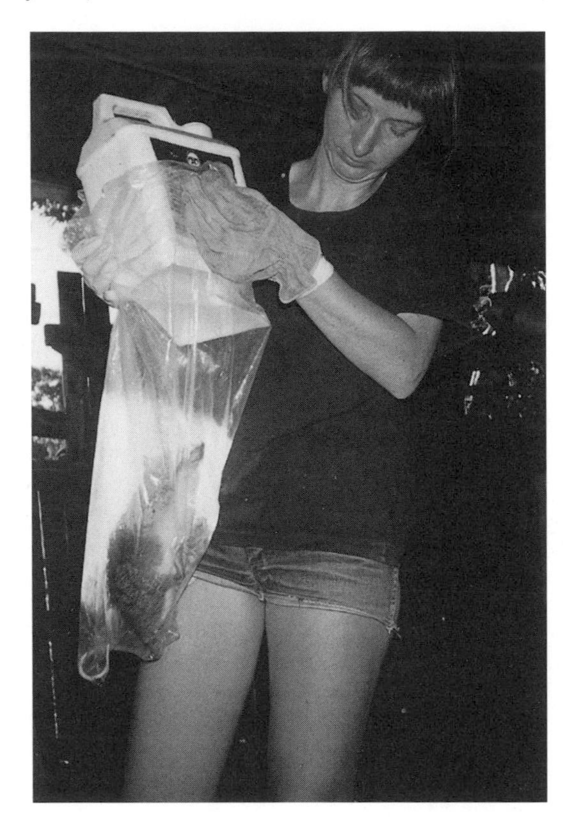

Chris Buehler gets ready to weigh a giant white-tailed rat in a study of rodent food preferences. The captive rats became amazingly tame around Chris—I'd never seen anything quite like it. PHOTO BY AUTHOR.

rainforest fruits and seeds to insects and earthworms. At the end of the day, Chris would take each rat out and weigh it, then weigh the amount of each food left over.

Like an ersatz Doctor Doolittle, Chris talked to the rats, making little squeaks and coos, and they became remarkably tame around her. She even had them trained to leave their nest boxes on command and climb into a little sack so she could weigh them. They'd take a chomp out of anyone else who tried to touch them but with Chris they were like kittens. I'd never seen anything quite like it.

But Chris's favorite pets were Reagan and Gorbachev, a pair of spectacled flying foxes. Flying foxes are so different from regular bats that, for many years, some scientists maintained that they were more closely related to primates than to other bats. A large flying fox can have a wingspan of four feet, and they have huge, intelligent eyes and a long snout, giving them a very foxlike countenance—hence their name. Unlike much smaller bats, which

A Boyd's forest dragon. This fellow, named Walter, turned up disoriented in a dairy shed. He became a permanent member of our household and loved to climb up the drapes. PHOTO BY AUTHOR.

navigate at night by echolocation (emitting high-frequency clicks, then listening for their echoes as they bounce off objects) and prey mostly on insects, flying foxes use their huge eyes and keen sense of smell to locate fruits and flower nectar.

Flying foxes are highly gregarious, living in colonies that once contained hundreds of thousands, perhaps even millions of animals. Australia's early settlers described how huge colonies of flying foxes could blot out the sun for minutes. Today, however, they are far less abundant, the victims of widespread forest clearing and persecution for their habit of occasionally raiding fruit orchards.

One of the fragments we were studying contained a colony of flying foxes, numbering perhaps one or two thousand animals. We'd trapped in the fragment several times and suddenly, in the late wet season, we began finding dozens of sick and dying foxes each day. We had no idea what was causing it. We took several of the ailing animals home and tried to nurse them back

A bumpy satin-ash, a rainforest tree that bears fruit and flowers on its trunk—adored by flying foxes and birds. PHOTO BY CHRISTINE BUEHLER.

to strength. They tamed down immediately, taking pieces of banana and mango from our hands. Although most died, two survived, and were christened Reagan and Gorbachev after the two world leaders.

To Reagan and Gorbachev, Chris was Mom. They were never happy unless sitting in her lap. Neither of them regained the ability to fly, but if Chris walked into the room they'd launch themselves at her and frantically flop-flop-flop across the floor, using their wings like oars. It was funny and pathetic all at once.

Later, in the dry season, the flying fox plague struck again, and this time it was even more alarming because many of the dead and dying were mothers with babies. The mothers would lie on the ground and flap helplessly while the babies screamed nearby. We tried to save several babies, feeding them milk from a tiny nipple and even taking them to bed at night to keep them warm, but were never able to raise one successfully.

The plague that had struck our colony was also devastating other colonies in the region. Eventually, a bat biologist by the name of Hugh Spencer figured out what was killing the flying foxes in such large numbers—paralysis ticks. Hugh believed that, because so much rainforest had been destroyed, the flying foxes were being forced to spend more time feeding near the ground, where they were likely to encounter the ticks. Flying foxes are especially fond of fruits from the wild tobacco tree, a shrubby, exotic species that proliferates in pastures. In fact, one night I found myself laughing at the spectacle of a flying fox and young coppery brushtail possum squabbling over a bunch of tobacco-tree fruit. The flying fox was hanging just above the fruit while the possum was just below, and the night was filled with their squawks and protestations, as neither was willing to give ground.

Unfortunately, when a flying fox was parasitized, the only way to save it was by injecting it with tick antivenin, which cost a hundred dollars a dose. This obviously was too expensive to save the thousands of poisoned flying foxes each year, but the babies were different. They rarely had ticks; they just fell from the trees as they clung to their mothers.

Once it became understood what was happening, animal lovers from all over north Queensland were galvanized to help raise the baby flying foxes. Local veterinarians learned to mimic flying fox milk, and suddenly the babies began surviving. Not that raising a flying fox was easy; the orphans screamed bloody murder if they weren't held constantly, and demanded frequent feedings. No one, it seems, could manage to raise more than one or two at a time; they were simply too much work.

The efforts to rear baby flying foxes soon became so sophisticated that large open pens were constructed near colonies as halfway houses. When they were several months old, the babies were released into the pens to slowly become assimilated into the colony, and if hungry they could always return to their pen and find fruit that was left there for them.

Though it was a wonderful scheme, it was a sad day indeed when all the moms and dads had to say goodbye to their babies as they were left, en masse, in the pens. Most of the waifs howled as they watched their tearful parents depart. Many were petrified of other flying foxes, especially if they'd been raised alone, having no idea what those other frantically screaming creatures were. Some of the babies desperately grasped little stuffed animals, the final legacy of their surrogate parents.

the world heritage controversy

Despite the miserable conditions of the wet season, Chris and Steve's brilliant organizational skills kept things moving along nicely. It was usually too wet at night to go spotlighting, though we still zipped out whenever the rain let up. The rainforest was a cacophony in the dark hours, the frogs in a frenzy, a dozen species trilling and squawking all at once. The leeches—true rain-lovers—were worse than ever.

After five or six hours of trapping each morning, we'd race home drenched and muddy, then peel off our filthy clothes and march straight into the shower. Our socks often bore silver-dollar-sized bloodstains because, no matter how much insect repellent we used, leeches managed to slip through and engorge themselves on our ankles or legs.

We were washing several loads of muddy clothes each day, until the motor on the washing machine eventually burned out. In the week it took us to get it replaced we washed all of our clothes by hand. There was no sense trying to dry anything outside on the clothesline, so the alcove over the wood-burning stove became festooned with complete wardrobes.

On most evenings the gang sat around and played cards or watched TV, while I stayed with Jenny in her small house just outside town. We'd learned to appreciate the simple pleasures of being warm and clean and safely indoors on these miserable nights, the house shuddering from the howling wind. We swore we'd never take being comfortable for granted again.

We were avoiding the Millaa pub these days. With increasingly determined calls from conservationists to designate the region's rainforests as a

World Heritage Area—an act that would close down the logging industry—the locals were growing very tense. I'd recently been shown on the front page of a regional newspaper, the *North Queensland Gazette*, holding a rare marsupial and advocating World Heritage listing. As a result of such activities our honeymoon with the locals—at least some locals—was clearly coming to an end. The leader of a logging crew, a red-haired chap named Paul, had been openly rude to us when we unexpectedly ran into him in one of our fragments. Most alarmingly, he shocked us with a thinly veiled threat. "A fella ought to watch his step in these parts," he said darkly, looking at me. "Things been known to happen to people out in the scrub."

Soon after that, Will Chaffee, who'd left the project in November but was living in a cabin owned by a farmer he'd befriended, had a bad experience in the Millaa pub. Will, as usual, pulled few punches in expressing his opinions, and was slammed up against a wall and screamed at by a farmer who'd taken umbrage when he dared defend World Heritage listing.

Few of the locals understood what World Heritage was really about. Crazy rumors were rife; a common proclamation was that World Heritage listing would allow communist countries like the Soviet Union to tell people in north Queensland what to do with their land. This was ludicrous, of course, but a surprising number of normally rational people took these arguments seriously. The owner of the Millaa gas station, an Armenian immigrant named Artie, was extremely agitated by the talk of communist takeovers. He told me—in utter seriousness—that he would sell his gas station and move away if World Heritage came to pass.

In reality, World Heritage is nothing like these paranoid rumors suggest. Australia, like the United States and nearly a hundred other countries (including the former Soviet Union—hence all the talk about communist takeovers), is a signatory to the World Heritage Convention, which is sponsored by the United Nations' Environment, Scientific, and Cultural Organization (UNESCO). The convention was enacted to promote the preservation of resources of unique natural, cultural, and historical significance. The Taj Mahal and Egyptian pyramids are well-known World Heritage sites.

In Australia, World Heritage listing had first been used in the early 1980s to stop the Franklin Dam, which would have flooded a vast, pristine area of Tasmanian forest. The dam controversy had created a bitter rift between conservationists and prodevelopment forces in Australia, and had helped launch the political career of Bob Brown, a courageous Green Party member of the Tasmanian Parliament who led the proconservation forces.

As the Franklin Dam issue demonstrated, what World Heritage listing did in fact do was to give the Australian federal government a lot more power to influence the management of a natural area. Unlike the United States, where the federal government controls and manages vast areas of public land (as national parks, national forests, Bureau of Land Management lands, etc.), the Australian Constitution grants the states free reign to develop their natural resources. Indeed, "national parks" in Australia are not "national" at all, but are actually controlled by the individual states. Some states—notably Queensland and Tasmania—were notoriously prodevelopment. Australia's federal government, however, is responsible for international treaties such as the World Heritage Convention, and the Franklin Dam set a great legal precedent. For the first time ever, the Australian High Court ruled that an international treaty could take precedence over a state's rights to control and develop its natural resources.

The Franklin Dam decision was an enormous blow to prodevelopment forces, and a boon to conservationists, for two reasons. First, local and state governments are almost always more vulnerable to development pressures than the federal government because they stand to reap the direct economic benefits. Second, people in cities are often more sympathetic to conservation initiatives than those living in rural areas. In Sydney and Melbourne, for example, there was massive support for rainforest conservation, and World Heritage provided a mechanism whereby the urban majority could impose its will on the rural minority. If the federal government heeded the calls of conservationists to designate north Queensland's rainforests as a World Heritage Area, logging would cease and management of the area for its natural values and tourism would become the top priorities.

A key event in the growing national debate was the publication of a book, *Tropical Rainforests of North Queensland: Their Conservation Significance*, by Aila Keto and Keith Scott, the former academics turned full-time conservationists whom I'd recently visited in Brisbane. Exhaustively documented, the book presented an overwhelming case for the global significance of north Queensland's rainforests. Prominent Australian biologists such as Len Webb, Peter Stanton, and John Winter threw in their support as well, while eminent British and American scientists like Norman Myers, Sir Peter Ashton, and Daniel Janzen visited Australia at Aila's behest and gave a series of hard-hitting lectures in support of World Heritage nomination.

As the controversy deepened in north Queensland, "World Heritage" became a local buzzword. It was often uttered with a disgusted, bitter expression, as though the person had just tasted something awful and needed to spit. For many, World Heritage became a litmus test. If you were for it, you were an enemy; against it, a friend.

This kind of black-and-white reckoning left precious little wiggling room for us. I'd been taking considerable pains to portray myself as a professional ecologist—not a rabid greenie. Though strongly advocating conservation, I tried to keep my arguments scientific and factual. Naively, perhaps, I thought we could live and work in Millaa Millaa while carrying on our conservation activities with the grudging acceptance of the locals. Increasingly, it seemed that this just wouldn't be possible.

Biologists sometimes need to be devious, especially if they work with covert creatures. The nocturnal mammals we censused by trapping, for example, posed quite a challenge. What were they eating? Where were they sleeping? How did they interact with other species in the forest?

The answers to these questions remained elusive—we simply couldn't see what the animals were doing. There is only so much that can be learned by catching an animal, tagging and releasing it, then catching it again. Christine's diet experiments with the rats in our garage were interesting, but the situation was so artificial that we really couldn't be sure whether these findings were valid.

We needed to see our critters in action, in the rainforest, doing what they normally do. We decided to try a technique that had been suggested by Greg Richards, a bat biologist in north Queensland. Greg had devised an unusual way to watch flying foxes at night.

Our Great Nocturnal Mammal-Watching Experiment actually began one morning, in our 200-acre fragment. Instead of releasing our animals as usual, we put each back into its own trap, then loaded the traps into our truck and headed home.

As evening approached, we launched into action. Steve pulled each rat out of its trap and Christine gave it a haircut. While this was going on, I shoved a syringe into a cyalume tube. Sold as highway emergency lights, cyalume tubes contain two chemicals that, when mixed by breaking an intervening seal, glow brightly for eight hours. I injected the glowing fluid into small gelatin capsules, the kind used to hold vitamins. One of these was then glued onto the back of each animal's head. We'd hand-colored the capsules with ink, so that our rats were now color-coded: the fawn-footed melomys yellow, the bush rats red, and the white-tailed rats green.

Thus, with two-dozen-odd brilliantly glowing rodents, we returned to the forest and set the animals free. Most began foraging almost immediately.

TOP Steve in cozy house slippers on the back porch of our Millaa Millaa farmhouse. PHOTO BY CHRISTINE BUEHLER.

BOTTOM Chris and Jeremy Blocke, an affable Englishman, contemplate a Millaa Millaa waterfall. The rubber boots—called "wellies"—were derigueur in the muddy rainforest. PHOTO BY STEVE COMPORT.

At this point we had nine people in our crew: all were positioned strategically around the forest and sat silently with pencils, notebooks, and red-filtered flashlights (invisible to nocturnal mammals) to record their observations.

In scientific terms, it was a wild night. Though the forest was pitch black, the lights worked beautifully, and many differences between the species became apparent. Bush rats were like erratic bulldozers, moving constantly across the forest floor, probably searching for insects in the leaf litter and rotting logs. The melomys spent most of their time in the trees, rarely ascending more than a dozen feet, though occasionally climbing down to feed on the ground. The big white-tailed rats often climbed right up into the forest canopy and could be heard gnawing at tough seeds like the Queensland walnut.

I'd often puzzled at how the cuddly melomys managed to coexist with the larger and highly territorial bush rats. A two-minute observation solved the mystery. A melomys had slowly descended a tree to the ground, where it was probably feeding on fallen fruit. Suddenly, the red light of a bush rat approached rapidly. The melomys shot up the trunk, where it remained, frozen, five feet above ground, as the bush rat bumbled by. After a few moments, the melomys slowly descended again to feed on its fallen fruit. Clearly, the melomys' superior climbing ability allowed them to thrive in bush-rat territory.

After a few hours the action slowed, and we packed up and went home. We were tired but ecstatic; in a single night we'd learned as much about the rainforest rodents as we had in the past several months.

Though our primary raison d'être was to study mammals, we were also learning something about the southern cassowary, one of Australia's most unique birds. A distant relative of emus, rheas, and ostriches, cassowaries stand over five feet tall and have bright red-and-blue necks and dark, loosely feathered bodies. Tall bony crests, termed casques, sit atop their heads. The casque's function is unknown—one theory holds that it helps the bird crash headfirst through thick vegetation.

Cassowaries feed on a wide array of rainforest seeds and fruits, which they gobble up from the forest floor. For several dozen kinds of large-seeded trees, cassowaries are the only known seed-dispersers, and thus could be crucial for the survival of those species. Biologists refer to such animals as "keystone species," the idea being that if they should become extinct, a number of other dependent species could disappear also (in the fabled Roman arch, the keystone is the central stone at the top, without which the arch would collapse).

Cassowaries have a very unusual mating system. Among most birds, either males and females pair with a single mate (monogamy), or males compete for mating opportunities with several females (polygyny). Cassowaries, however, have reversed things. Female cassowaries mate with a series of males in succession (polyandry), and each male ends up brooding his own eggs and raising his young. No one knows why this odd system evolved—only a handful of birds in the entire world are polyandrous.

For such an imposing bird, cassowaries can be surprisingly secretive. We were, however, able to assess their abundance in different areas by searching for their very distinctive dung, which consists of a big pile of semidigested seeds. Altogether, we found sixteen cassowary dung piles in our control areas in continuous forest, and none in any of our fragments.

According to the locals, cassowaries had been shot out of most fragments (apparently for sport), or killed or driven away by dogs. Because it requires extensive areas to live in, there is real concern among biologists about the cassowary's ability to survive in Australia (the species also occurs in New Guinea, along with two other cassowary species). Strictly confined to tropical rainforests and coastal swamps, southern cassowary populations have been reduced and fragmented by massive forest clearing, especially in the coastal lowlands. At Mission Beach south of Cairns, one of the last strongholds of the species, starving cassowaries had to be fed by National Parks rangers and residents in 1986 after a cyclone decimated the area's forests. Prior to forest clearing, the cassowaries at Mission Beach probably would have just migrated elsewhere to find food.

Though shy, cassowaries are also unpredictable. On rare occasions they've attacked people. With massively strong legs and daggerlike claws, an upset cassowary is not to be taken lightly. While no one really knows why cassowaries suddenly attack, Les Moore, an Australian biologist who studied them for years, believes they particularly dislike joggers—the regular thumping sound seems to set them off. At other times they've attacked for no apparent reason. A middle-aged man, for example, was savaged by a cassowary while simply out walking with his wife. The man nearly died after the cassowary kicked a five-inch-deep gash in his chest.

Despite such risks, some biologists have spent years in close contact with cassowaries. Several cassowaries at Mission Beach habituated to Joan Bentruppenbaumer, who followed them around while studying their behavior, the birds carrying on their normal lives as though she didn't exist. Les Moore also studied cassowaries in the field. One bird, he recalled, lay down to rest while he was watching it. Feeling tired too, Les propped himself against a tree a dozen yards away and soon drifted off. When he awoke, he was shocked to find that the amiable cassowary had snuck over and gone to sleep right next to him.

joh starts his run

Somewhat to my surprise, Shirley Taylor's initial interest in conservation was growing into a serious, full-fledged commitment. She dropped by Jenny's house every day, and she and I'd get so immersed in rainforest politics that we'd sometimes forget Jenny was there. She'd been sickened, she said, by all the forest clearing she'd witnessed when she was young; now it was time to stop. She'd begun attending local meetings of the Australian Conservation Federation, and was pumping me for information. A conservationist had been born, and come hell or high water, Shirley Taylor was going to fight in the trenches for the forests.

Shirley's transformation was all the more amazing given the profound social risks she was taking. The tide of local opinion at this point was swinging massively against World Heritage listing. By throwing her lot in with the conservationists, Shirley was defying the neighbors and friends she'd spent the last five decades with. The pressure to conform was enormous, and it was hard not to feel anxious for her. I'd weathered these kinds of battles before in the United States, but I wasn't sure whether Shirley realized what she might be getting herself into.

Though Shirley was warm and open, Jenny's father and brother weren't quite as easy to befriend. Both were Queensland country boys, and we struggled to find common ground—aside from sports, the universal bloke language. The good will was certainly there, but when left alone we often lapsed into awkward silences. This gulf was brought home to me one day when Jenny's former boyfriend, Ken, paid a visit to the Taylor's farm to see his daughter, Kasie. A country boy himself, Ken struck up a rollicking conversation with Jenny's dad and brother, and they all carried on like great old mates. Fortunately, I found Ken to be a very decent guy, and if his visit sparked a twinge of jealousy in me it seemed doubly awkward for him. After five years together, Ken had wanted to marry Jenny and had been devastated when she'd refused.

Kasie, who was now speaking a few words and toddling around the house, was an equally tough nut to crack. I'd spent very little time around babies, and initially felt quite awkward around her. Today I've learned that if you just ignore kids, they'll soon be climbing all over you, but back then I wasn't quite sure what to do.

For her part, Kasie tolerated me fine so long as Jenny was present. As a single mother, however, Jenny rarely got a break, and she really got worn down some days. At night we'd often drive around in my truck because the engine's droning sound lulled Kasie to sleep.

One morning in March as the sun finally broke through the clouds, Jenny realized she was out of cigarettes. I offered to drive to the Millaa store, a quarter mile away, but she wanted to walk. I could stay and keep an eye on Kasie, she said.

"But, but . . . " I sputtered. "She won't like that at all, you know. Not one bit."

"She'll be all right," Jenny said confidently. "I'll just sneak out and be back before she knows I'm gone."

So Jenny slipped out. For a few minutes Kasie played quietly in the bedroom and I sat in the living room reading.

Then Kasie came out and said "Ma . . . ?" Cringing, I figured my only option was to bluff: I began chatting away casually, as though Jenny were just in the kitchen. This trick worked for about ten seconds, but then Kasie toddled into the kitchen, took one look around, and let out a bloodcurdling howl that would've freaked out a werewolf.

The next fifteen minutes seemed an eternity. Kasie screamed hysterically, beet red and barely stopping to breathe, while running frantically from one room to the next as fast as her wobbly legs would carry her, desperately searching for Jenny. Nothing would console her; if anything she just screamed louder when I tried to calm her down.

By the time Jenny returned, Kasie's and my nerves were shattered. On the verge of panic myself, I kept thinking someone would drop by and think I was beating the poor waif. Kasie climbed into Jenny's lap, sniffled a few times, and fell sound asleep. As I stood there quivering, one thought kept popping into my head: Who's supposed to calm *me* down?

As if the growing World Heritage debate weren't bad enough, we learned in March that Queensland's premier Joh Bjelke-Petersen had decided to run for prime minister. The thought of him as national leader made us all cringe. Long a cunning advocate of state's rights—the notion that the federal government should butt out of virtually everything except foreign affairs—Bjelke-Petersen was now threatening to make the World Heritage listing of Queensland's rainforests an even bigger political hot potato than it already was.

Bjelke-Petersen's run for PM was wildly popular in rural Queensland and Tasmania. Elsewhere in Australia, however—especially in southern cities like Sydney and Melbourne—he was perceived as a dangerous kook. In interviews, the sophisticated southern media peppered Bjelke-Petersen with tough questions, and his arrogant retort—"Don't you worry about that!"—rarely failed to make the evening news.

The Bjelke-Petersen run only served to heighten the long-standing tension between Queensland and its southern neighbor, New South Wales. In the eyes of Queenslanders, New South Welshmen—derisively called "cockroaches"—are haughty and conceited, especially if they live in Sydney, which brashly bills itself as Australia's leading city. To New South Welshmen, Queenslanders—lowly "cane toads"—are a bunch of rural rednecks. Being a sunbelt state with a growing economy and population, however, Queensland is fast becoming a dynamic force in Australia's social milieu.

The deep antipathy between cockroaches and cane toads comes to a dramatic head each year in three shattering collisions termed the State of Origin Competition. These are, without exaggeration, the most violently contested rugby-league matches on the planet. For someone who grew up playing American football, to watch a State of Origin match is to marvel that players could be hit that hard, that often—without helmets or pads—and still walk off the field in one piece.

Because of such stormy undercurrents, the forthcoming contest between Bjelke-Petersen and Bob Hawke—the gregarious, backslapping Sydney pol and former union leader who was the current PM—was shaping up as much more than a mere political race; it was an epic battle of left versus right for the heart and soul of Australia. Seeing the political opportunity, Bob Hawke soon declared his intention to forge ahead with the World Heritage listing of north Queensland's rainforests—regardless of what the intransigent Queenslanders thought—if reelected prime minister. North Queensland's rainforests had just become a crucial political football. Their fate would be decided in the approaching election. The battle lines had been drawn.

Their country boasting the world's deadliest snake fauna, many Australians abide by the maxim that the only good snake is a dead snake. It matters little if it is a fearsome taipan or harmless python, if it comes into contact with humans it usually ends up dead.

As biologists, however, we considered ourselves more discriminating. We didn't kill *any* snakes, and we took pains to show no fear of the harmless species. If we happened to see a carpet python or amesthytine python in the field, we'd pick it up to show everyone its interesting features—like the heat-sensitive baffles on its jaw for detecting warm-blooded prey at night—while wrestling with its powerful coils. I'm no big fan of snakes, mind you, but I've managed to overcome my fear to the point that I'm only mildly nervous around pythons and the like.

Another quite harmless species is the brown treesnake. Although fairly big (up to ten feet long) and capable of being aggressive, it is only mildly venomous and no real threat to humans. The only problem with brown treesnakes is that they look a lot like brown snakes, which are *highly* venomous and must be treated with great respect. Aside from subtle anatomical features of the head, the main difference between the two species is that brown snakes rarely climb trees.

It was because of this difference that I agreed to a request from Erv Petersen at the School for Field Studies to remove a big brownish snake that had taken up residence in their house. We'd been visiting the SFS students when the snake was discovered. No one showed much interest in grappling with it, as it was perhaps seven feet long and elaborately wrapped around the exposed wooden rafters just above the female students' bunk beds. Les Moore was also visiting that day and so, after egging each other on a bit, we confidently volunteered to solve their snake problem.

After a brief consultation, Les and I agreed that it must indeed be a brown treesnake. Though we couldn't see its head, it was obviously a good climber, so the odds weighed heavily in our favor. With that reassuring thought in mind I climbed atop a table and began grappling with the snake, which was surprisingly strong and utterly determined to stay right where it was. A crowd soon formed around us, with the entire group of SFS students and staff plus my volunteers all clustering around to watch the excitement.

Knowing that brown treesnakes can be vicious biters, I'd begun with the snake's tail, and was slowly disentangling it from the rafters. Les kept glancing up nervously, trying to get a look at its head—brown treesnakes have more protruding eyes than the dangerous brown snakes. Les's jumpy demeanor wasn't helping me, given that only moments before he'd confidently agreed with my suggestion that it must be a treesnake. The snake and I were in a major tug-of-war, and I'd just started making real progress when Les shouted, "It's a brown snake!"

I dove off the table, landing on top of several students who were desperately trying to scramble out of the way. Somebody screamed, fueling the panic. In three seconds we'd all stampeded to the far corners of the room. Many students were cringing in fear, as though the snake might suddenly dive down upon us in an aerial assault.

I shot a glance at Les, who suddenly looked quite sheepish. Normally I have great confidence in his ability to identify wild animals, but his body language told me he wasn't at all sure what he'd seen.

Les and I had another little powwow. Neither of us had gotten a good look at the snake's head, but the seeds of doubt had been sown. While the odds were still in favor of a treesnake, my enthusiasm for bare-handedly dragging it down from the rafters was now greatly diminished. Les declined my offer to stand aside and let him demonstrate his snake-wrestling prowess, so we collectively decided just to leave the snake alone. No doubt it'd leave on its own, eventually.

Our pronouncement that the snake would stay was met by glum silence. Until then, the students had seemed rather impressed by our expertise and bravado. Especially unhappy were all the coeds who now had to face the prospect of sleeping with a large and possibly dangerous snake coiled up in the rafters just above their heads.

Fortunately, the following day Erv phoned to tell me the snake had disappeared during the night. No one had slept in that room. Les and I continued to visit the SFS gang, thankful that their snake problem had resolved itself. Though we were no longer fearless Crocodile Dundees in their eyes, at least we weren't complete goats either.

right in the thick of it

As the wet season drizzled to a close the project was zipping along beautifully. To keep everything running, Chris, Steve, and I usually worked seventy-hour weeks. When Steve appeared frazzled I'd give him fifty dollars and tell him to go skydiving. He'd return looking like a new man. Chris seemed rejuvenated just by being in the forest. The only time she looked out of sorts was when she got a leech in her eye and we had to wait until we got home to extract it with a pair of tweezers.

Volunteers came and went. Some were unforgettable—from Tim and Jennifer, a zany British couple who grew more British by the minute; to Sam, a laconic Aussie triathlete who did a full day's work then shot off for ten-mile runs while the rest of us collapsed in exhaustion; to Dan, a California surfer who called everybody "dude."

After nearly a year of fieldwork, it was clear that the forest fragments we studied were being drastically altered. Many animals had declined sharply or disappeared, such as the cassowary, lemuroid ringtail possum, musky rat-kangaroo, and Atherton antechinus, to name but a few. Assemblages of native predators were also disrupted; larger predators such as spotted-tailed quolls, pythons, and rufous owls had disappeared and were replaced by small, opportunistic species like white-tailed rats and boobook owls. Fragments were especially vulnerable to damage from windstorms, which toppled trees and ripped holes in the forest canopy, allowing light-loving species like wait-a-whiles and stinging trees to flourish. Exotic weeds and poisonous cane toads were invading the fragments from the surrounding pastures.

While pleased with the project, we were struggling with an important decision: whether to continue our involvement in the World Heritage debate or, as one logger bluntly told us, "Just pull your bloody heads in!" While we all wanted to carry on, it was a very tough call. Aside from the growing risk of social ostracism in Millaa Millaa, there was a very real chance that the farmers whose land we relied upon would get fed up and kick us off. This would be disastrous, as we still needed a lot more information on quolls, antechinuses, and other rare or elusive species.

We were also getting friction from unexpected quarters. One day I received a phone call from a prominent (but conservative) Queensland biologist. In an agitated manner, he demanded to speak to me immediately "about how my conservation activities were detracting from my research." I met with him the next day, and he explained that he thought it very unseemly for scientists to become embroiled in politics—it brought their impartiality and objectivity into question. He also said he'd been receiving complaints and queries about me from some very high quarters in the Queensland government. I doubt that he was expecting my reaction, however, which was undisguised delight at the thought that I'd struck such an important nerve.

But after weighing up all the arguments and risks, we just couldn't get over the thought that we could be badgered into not doing something we really believed in. It was clear, moreover, that we'd been having a positive role in the debate—a small one, no doubt, but we were right in the epicenter of the conflict and were making ourselves heard. We had a conference and the feelings were unanimous: we'd continue our conservation activities—and damn the torpedoes.

With the federal election looming, there was a maelstrom of activity surrounding World Heritage. The towns on the Atherton Tableland—especially the nearby logging town of Ravenshoe, the heart of anti–World Heritage sentiment—were being aggressively lobbied by conservative politicians. Facts were optional. Bob Katter, the minister for northern development, was especially alarmist. "All further development would be stopped in its tracks," he proclaimed, if World Heritage came to pass. The Queensland forestry minister, Bill Glasson, claimed that 3,500 families would be homeless as a result of the listing. Joh Bjelke-Petersen said that the federal government didn't realize that Queensland was already looking after the rainforests and that they "could go jump in a lake."

Despite the dire predictions of right-wing pols, a rational look at the situation suggested that as few as 350 jobs would be affected. Moreover, the federal government had committed upward of fifty million dollars for various programs to offset these job losses. Displaced timber cutters would be offered jobs in reforestation and plantation programs to compensate for the reduction in timber harvests. Because most of the plantations would grow softwoods like Caribbean pine, rather than tropical hardwoods, the government would pay to have the timber mills retooled to process these timbers. Limited harvests of tropical timbers for local crafts industries could continue on private lands.

That is not to imply that the changes wrought by World Heritage would be trivial. Perhaps the greatest impact was psychological; the timber workers were a fiercely independent breed, and the last thing in the world they wanted was to be employed on some government-sponsored make-work program. They saw themselves as pioneers—or at least the direct descendants of pioneers—who in their own way were effective stewards of the rainforest. I saw a television interview of the wife of a Ravenshoe timber worker. The interviewer asked the woman what kind of work they'd be willing to do if World Heritage came to pass. The woman was adamant—they only knew how to cut timber. That's all they would *ever* do—cut timber.

It is true that the logging operations managed by the Queensland Forestry Department were superior in many ways to the rampant cuts going on today in places like Southeast Asia, New Guinea, and the Amazon. Under increasing pressure from conservationists, Queensland had developed a series of guidelines designed to reduce the impact of logging operations.

Ecological issues aside, however, there were two serious problems with this system. First, it was quite expensive to run; the government covered the costs of the Queensland Forestry Department and its professional field officers, who oversaw the logging operations, and also paid for the construction of some logging roads. Ultimately these costs were borne by taxpayers, who were effectively subsidizing rainforest logging whether they liked it or not.

Second, the timber harvests were steadily declining as a result of past overcutting. Most tropical timbers grow slowly and thus long intervals between logging cycles—perhaps a century or more—are needed for timber stocks to fully recover. Indeed, some big trees removed during logging are ancient—up to 1,400 years old, based on recent research in the Amazon. Moreover, because the area of accessible rainforest in north Queensland is very limited, the timber industry had always depended on logging some virgin forest each year in order to maintain its harvests. As time went on, virgin forests became harder and harder to find. Annual timber harvests had declined from over 200,000 cubic meters in the 1970s to only 60,000 cubic meters by the mid-1980s, and most projections suggested that further reductions would be needed. The concept of a "sustained yield" long advocated by Queensland politicians was in fact a myth. Even without World Heritage, the industry was dying.

In the ongoing debate, the Queensland Forestry Department was assuming a tack that I found highly peculiar. Having participated in similar battles in the United States, I was accustomed to a situation in which resource-management agencies such as the Forest Service and Bureau of Land Management played middleman—beleaguered and nominally neutral bureaucrats trying to find the middle ground while conservationists and prodevelopment interests fought it out on either side of them. This was the natural order of things—or so I thought.

In Queensland, the Forestry Department had abandoned any pretence of neutrality to become a full-blown opponent of World Heritage. This probably reflected the heavy politicization of the department, and the fact that it was a state, not federal, entity. Hand-in-hand with the loggers—the group they were supposed to be regulating—the QFD began a series of "public information meetings" in towns around the Atherton Tableland.

When I found out about the meetings, I visited the regional forester in Atherton, an amiable fellow named Ted Mannion, to try to get myself included in the program of speakers. I also phoned the wife of a timber worker in Ravenshoe who was organizing the meetings. I ran into brick walls both times. If they were really public information meetings, then shouldn't someone be allowed to present a responsible opposing viewpoint? No, I was told, only representatives of the QFD and logging industries were invited to speak. Despite the veneer of providing a public service, I realized, there would be very little pretence of balance.

The first of these meetings was held in a vast Quonset hut that served as the public hall in Malanda, twenty minutes from Millaa Millaa. The place was crammed full; perhaps five hundred people attended, of which 99 percent were anti–World Heritage. The only proconservation forces were myself and three biology students from James Cook University, though a few others may have elected to remain incognito.

The meeting was raucous at times. One expected the logging-industry representatives to act like cheerleaders, but the thing that really made my blood boil was the dogmatic "scientific management" tack taken by the QFD representative: they were doing a wonderful job, he claimed, and here were all the statistics and graphs to prove it. I could see a half dozen gaping holes in his argument, but the audience lapped up his proclamations like they'd been delivered by Moses himself.

Finally, there was an opportunity for people in the audience to ask questions. After a few moments I raised my hand, feeling butterflies dance in my stomach. I began by explaining that I was a biologist doing research for my doctorate in the tableland's rainforests. This was a good way to start, I'd learned; regardless of their political persuasion, most people respected the views of a scientist. Then I explained that the speakers had been somewhat unbalanced in their presentations—several faces fell—and began counting off the most striking discrepancies, one by one, my momentum building as I went. I finished by reminding the audience that the rainforests of north Queensland were among the most ancient on the planet, and were truly of international significance.

My heart was thudding when I sat down. My little speech had lasted perhaps two minutes, but I'd covered a lot of ground. I waited for a reaction from the huge audience—Catcalls? Boos? But there was nothing, just a grudging acknowledgment and a few nodding heads. I'd been serious and credible when I'd spoken, so at least it was now clear that there was a legitimate viewpoint that wasn't being fairly represented here.

A few minutes later, one of the James Cook University students stood up as well. He said he thought that a limited amount of logging would be acceptable if bulldozers and other heavy machinery were banned from the forest. Rather, loggers should go back to using bullock teams and wagons to haul the logs out of the forest, the way they'd done at the turn of the century. Once the audience realized what he was saying, the jeers started and he was quickly drowned out by boos.

Two days after the Malanda meeting Steve found a note stuck under our windshield wiper. It was addressed to "Willy Longknife In Our Back." In a scrawling hand, it abused me for violating the trust of the people of Millaa Millaa, for accepting everyone's generosity and then doing my utmost to betray them. It ended with a bald threat: If I didn't back off I'd soon get what was coming to me.

A few days later, Helen at the takeaway let me know that a former logger was telling everyone within earshot that I was a world-class ratbag, a lying backstabber, and that I tortured the animals I was catching. I decided to deal with this head-on. I drove over to his house and rapped on his open front door. I'd just wanted to talk, but when he saw me he almost fainted—like he thought I was going to kill him. Despite my assurances of peaceful intent, he refused to come outside, refused even to come into his living room where we could at least talk across the doorway.

But the incident that really burned me was finding out that my eccentric mate, Geoff Downey, had been humiliated by some neighboring farmers. Geoff was known to be sympathetic to conservation, and when a few of his cows escaped one of his neighbors refused to return them. Geoff drove over to talk to him and found a small gang of men waiting for him. According to Geoff, they sneered and laughed at him, implying that he was an unwanted oddball, and essentially made him beg to get his cows back. With his missing leg, Geoff could hardly stand up to them physically.

Shirley Taylor was also catching hell. Women she'd known for decades had turned bitter and cold, and some even abused her to her face. She was in shock, dark circles growing daily beneath her eyes. Shirley was a generous and sensitive woman, and I doubt she'd ever imagined that people could turn on her as harshly as they had.

I was seeing a kind of vindictiveness—even plain old meanness—that I hadn't experienced in earlier conservation battles. I suppose I shouldn't have been so surprised; when people believe their livelihoods are threatened, they will fight viciously. The loggers and farmers couldn't compete with our scientific arguments, so they fought with what they did have—their mouths and their fists.

The repeated physical threats—real and implied—were starting to get to me. After Will Chaffee's pub brawl and my "Willy Longknife" letter, I wouldn't have been at all surprised to walk around a corner and find a mob of angry loggers waiting to knock some sense into me.

Perhaps it was time for a little psychological warfare. I'd studied martial arts for a few years, and went home and donned a black T-shirt emblazoned with a giant fist and the words "Kuk Sul Won"—a Korean martial art—in bright gold ink. I strolled down to the Millaa pub, ordered a six-pack of beer, casually pivoted around so everyone got a good look, and left. I felt positively silly, but it was the kind of display that tended to resonate in a place like Millaa Millaa. At least it might make the locals think twice before trying to use me as a punching bag.

the rat trials

In early July Will Chaffee dropped by the house, and I couldn't believe how skinny he looked—he must've dropped twenty pounds. His gaunt face was wind- and sunburned. He'd been traveling, he explained, and had come very close to losing his life.

Ever searching for adventure, Will had put his head together with his old mate and cofounder of the Higher-Mammal Crew, Jeff, who was living in Malanda these days, and decided to head off in a search for one of Australia's rarest snakes, the Kimberlies python. Only three of the snakes had ever been seen by scientists, for it lives in one of the most remote wildernesses on earth—the Kimberlies Mountains of northwestern Australia.

With his usual verve, Will had written to the Australian Geographic Society, which publishes a national magazine, and asked them to underwrite the costs of their expedition. To his surprise, they agreed, on the understanding that he would contribute an article describing their adventures.

The Kimberlies are a vast, wind-carved range of ancient massifs and bizarre geological formations—such as the Bungle Bungles, which look like giant beehives. To reach the deep interior of the range, Will and Jeff had to be flown in by helicopter. They arranged to be picked up a fortnight later. Carrying with them everything they would need—medicine, food, camping gear, cameras, and maps—they set off in search of their elusive snake.

After two days of hiking and exploring, they came to a small river swollen dramatically from a recent rainstorm. They needed to get to the other side, so they waded in, using walking sticks for balance. The current was far stronger than they'd realized. Jeff was knocked over first, then Will followed. They were being pulled downstream by the torrent and dragged under by their heavy backpacks. Panic set in, and they lost both backpacks in a mad scramble to get ashore. They crawled out of the water, glad for the moment just to be alive.

But then it hit them—they had twelve days to survive until the helicopter returned, and they'd just lost everything. A very serious jam. A search downstream revealed no sign of their missing food and gear, and even an Aborigine would have been hard-pressed to survive for long in those desolate mountains.

They didn't even have matches, though they'd managed to hang on to one of their canteens. After much thought and discussion, they decided they had but one option—to set off for a small shack they'd seen in the distance the previous day.

After an exhausting hike, they arrived at the dilapidated shack, which was long abandoned and had probably belonged to a gold prospector. A search turned up a box of matches and a tin of pancake mix. Fortunately there was a small stream nearby, or else they wouldn't have been able to stay there.

So they made themselves at home, insofar as possible, and set their sights on surviving for a dozen days. Their daily ration was one small pancake each, cooked on a sheet of aluminum over a campfire. They had nothing to read, nothing to do but lie about and conserve their energy. Any thought of searching for their rare python had long since been abandoned. The nights were chilly and miserable, the days hot and stultifying. Without insect repellent, the flies were horrible. Fortunately, both of the young men had begun the trip with a little extra padding, fat reserves to burn.

The twelfth day finally arrived, and the long hike to reach their rendezvous with the helicopter did them in. But they did make it, and the helicopter arrived right on time, the pilot shocked to see them both so scrawny and bedraggled.

The pilot lent them some money, and after a few days of resting and gorging themselves in a small-town hotel, they felt well enough to make their way back to north Queensland. They hadn't found fame, or fortune, or the Kimberlies python, but they were happy just to be alive.

❧

To everyone's surprise, Australia's political scene changed dramatically in the few weeks before the federal election. Joh Bjelke-Petersen—who'd been running as an independent candidate, much like Ross Perot did in the United States in 1992—found that his campaign had faltered and collapsed. Moreover, Bjelke-Petersen's blunt, contentious style had divided Australia's conservatives, some of whom backed him while others supported John Howard, a far more moderate candidate. With the conservatives politically fragmented, Bob Hawke appeared to stand an excellent chance of reelection.

With the election all but decided, the prospects for World Heritage listing suddenly seemed far more likely. In desperation, conservative pols pulled out all the stops. In describing the UNESCO staff who administered the World Heritage program, Geoff Muntz, Queensland's minister for environment, conservation and tourism, resorted to blatant scare-mongering. "Some of the people put there to make these decisions are nothing more than puppets of communist and other extreme socialist regimes," he said. "They are more interested in political propaganda than conservation."

The agitation in Millaa Millaa was getting worse. I was especially concerned about Shirley Taylor, who seemed to age each time I saw her. I felt guilty for letting the Taylor family get dragged into this whole mess. They were all going to the wall for me—even Jenny's normally laconic father, who shocked everyone in the Millaa pub when he snapped back at someone who was disparaging me.

The truth is that I was stunned at how bitterly World Heritage was being contested. I'd expected people to be unhappy, but nothing like what I was seeing—especially the vindictive, personal attacks. Like the Millaa residents, my upbringing was rural—my grandfather was a farmer, my uncle a logging-truck driver, my dad a rancher—but I'd never seen behavior like this. I was reminded of a story I'd been told—a true one—about a young fellow who'd been caught buggering a cow in Millaa Millaa a few years earlier. The dairy farmer who caught him dobbed him in to the police, and in the ensuing scandal the young man was completely ostracized by the community. His family was so shamed that they had to move away. That was how we were being treated—like we'd just committed an unspeakable act.

In fairness to the community of Millaa Millaa, I have to emphasize that not everyone acted so boorishly. It's just that the really aggressive ones tend to make a lasting impression. I recall that during the darkest time of the controversy, there was a knock at our door. It was one of our young neighbors, Nicole, and she'd come to deliver a cheesecake that her mother had just baked.

On top of all the World Heritage dramas, our project was running seriously short of cash. We'd nearly exhausted the grants I'd received from the U.S. National Science Foundation, New York Zoological Society, and other organizations. If something didn't happen soon, we'd have to shut everything down.

A few months earlier we'd made some money for the project by acting as guides for a Japanese film crew, shooting footage for a wildlife show called *Wakka Wakka Animal World* (Wonderful Wonderful Animal World). Over the course of a week we'd taken them to our best spotlighting sites, and they'd gotten great footage of possums and tree-kangaroos. We'd been pleasantly surprised when they gave us a two-hundred-dollar bonus.

But now, we desperately needed another infusion of funds. I was thus delighted when Erv Petersen asked if I'd be interested in teaching an SFS course on rainforest ecology. Though only a month long, it was a full-time job, and I couldn't have taught and kept the project running without Chris and Steve. When the course started, they drove out to SFS every day to tell me about their trapping and spotlighting results and to talk over anything unusual.

We had only eight students—all undergraduates from the United States—and Erv and I ran the course together. He taught natural-resource management while I gave the students a good dose of rainforest ecology. We went into the forest almost daily—spotlighting, birdwatching, catching insects. We hiked up a mountainside, visited our fragments, and rode horses on Pat Reynolds's farm.

One of our more entertaining experiences was a day at the Millaa Millaa grade school. Most adults in the area were dead-set against World Heritage, but we were hoping the school kids would be more open-minded. Dressing up as cassowaries and tree-kangaroos, the SFS students performed several funny skits that were like ecological morality plays. To our delight, the kids were enthralled and roared at all the jokes. More important, they seemed to absorb our messages about the irreplaceable value of rainforests. The afternoon didn't go quite so well, however. The kids thrashed us in a game of touch-rugby—not only were we outnumbered but even the ten-year-olds had better rugby skills than we did.

Later in the course, Erv arranged for the SFS students to see a local logging operation. I ended up taking the group, and unfortunately things got very tense. As we toured the cutting area, accompanied by a young officer from the Queensland Forestry Department, I snapped photos of the loggers in action—not with any malicious intent, but simply to get some good slides of logging operations. The loggers certainly knew who I was, and when it was time for us to leave they demanded that the forestry officer confiscate my film.

This photo got me in very hot water with the Queensland logging crew, who tried to confiscate my film. PHOTO BY AUTHOR.

I was damned if I was going to give up my film—why should I?—but they were equally adamant. The young forestry officer was the proverbial meat in the sandwich, and had no idea what to do. He kept walking over to the bristling loggers, conferring intensely, then returning to plead with me to give up my film, just to keep the peace.

As a scientist and invited guest—on public land—I had every legal right to be there and take photos. This whole thing was pure intimidation, and I was determined not to cave in. But the loggers refused to let us leave. An experienced forester probably would have told them to go to hell, but our young officer was buffaloed. He kept appealing to me to help him resolve the situation.

Finally I relented, mainly because the students were tired and wanted to go home. I made the officer swear that nothing would happen to my film, and to promise that I'd get it back.

The next day I got a call from the district forester, Ted Mannion, who was in charge of the Atherton region. He was profusely apologetic, and asked me to come and get my film right away, even offering to pay to have it developed. When I picked up the film he thanked me for being reasonable. Under the same circumstances, he said, some unscrupulous greenies would have contacted the media and blown the incident up into a nasty scandal. The thought had never occurred to me.

The SFS course drew to a good close, and three of the students ended up joining our project. As a requirement of the course, each student had conducted a brief research project, and these three young women had teamed up to study the behavior of native rodents in our Millaa Millaa garage. These "rat trials" became rather infamous, and the results were so interesting that I invited the three to join our team so they could continue their work.

The students had constructed a circular arena, about eight feet in diameter, into which they'd introduce two rats simultaneously. The trials were run at night, when the rats were wide awake. The only light came from a dim red bulb overhead, which is invisible to nocturnal mammals.

The goal of the experiments was to determine whether some rodent species were competitively superior to others, and whether factors like rodent age and sex made a difference as well. An ancillary benefit was simply seeing the behavior of the animals in these tense, competitive situations. This was all new stuff, virtually unknown to science.

The trials were fascinating. The students would put two rats into opposite sides of the arena, and for a moment they would wander around quietly. Then they would see each other and freeze—their eyes bulging and every hair standing on end. Sometimes one animal would emit an angry whine.

Invariably, then, one of the rats attacked. The bush rats and Cape York rats—two closely related species—had a similar style: they'd turn sideways and scramble toward their opponents, slashing out with their hind feet at the other's face. The stylized attack often degenerated into a wild scramble, the rats moving in such a blur that the students couldn't take notes fast enough. Then, as quickly as it had started, it was over—one rat would concede and dash off, with the other in hot pursuit. In the forest, of course, the loser would vacate the contested territory, but in our arena they just ran around in circles. We quickly realized that we had to stop the trial at that point, or they'd run themselves into exhaustion.

The rat trials did more than reveal intriguing behaviors; they also gave us insights into the responses of these species to forest fragmentation. For example, being so closely related, the bush rat and Cape York rat have many behavioral similarities. Both forage on the rainforest floor, feed on insects and fungi, and vigorously defend their territories with similar styles of attack. For this reason, competition seems especially intense between these two. In large forest tracts, there is enough space for them to largely avoid each other, but in the confined universe of a forest fragment, coexistence is far more difficult. Our trapping studies had revealed that the two rats had a "checkerboard distribution"—most fragments sustained one species or the other, but rarely both.

These observations suggest one of the mechanisms that can cause species to disappear from fragments. Because of competition for limited space, food, and other resources, assemblages of closely related (and therefore ecologically similar) species may often collapse in fragmented forests. The animal communities become simpler and poorer, with intensified competition becoming a powerful driving force of extinction.

It was perhaps not surprising that the bush rats and Cape York rats both dominated the cuddly-looking melomys. Though pugnacious defenders, the melomys never initiated attacks, but leapt at the other rodents with a shrill squeak if approached—all bark and no bite. In fact, if we put a branch in the arena, the melomys just scrambled up to the highest point and stayed there, effectively ending the contest. Because they were far superior climbers, the melomys easily coexisted with the bush rat and Cape York rat—even in the smallest fragments.

Despite the fact that we lacked a real laboratory, we learned a lot from the rat trials. We realized that each species had a broad repertoire of interesting behaviors, that their interactions were often stylized bluffs. We also learned that an animal's past experiences were important; if it had lost its last encounter, it was likely to lose the next one too, even if paired against a smaller opponent that it would normally have dominated. Out in the rainforest there were wild little wars going on every night, contests with sophisticated rules that only the participants fully understood.

TOP A small rainforest fragment in the Millaa Millaa area. Note the forest is scrubby and short—the result of frequent wind damage. PHOTO BY AUTHOR.

BOTTOM Zillie Falls, northeast of Millaa Millaa. The waterfall was in a twenty-acre forest fragment, protected as a scenic reserve. PHOTO BY AUTHOR.

TOP A tree fern, an archaic plant that heralds back to the age of dinosaurs, witnesses a north Queensland sunset. PHOTO BY AUTHOR.

BOTTOM Some rainforest trees, like this stately giant, may be over 1,400 years old. PHOTO BY AUTHOR.

TOP Lantana, an exotic shrub from Central and South America. Although pretty, it is a noxious weed that invades rainforest, especially in fragmented or logged areas. PHOTO BY AUTHOR.

BOTTOM A brown treesnake. Notoriously aggressive, a snake like this gave me a big fright when I tried to drag it down from the rafters at the School for Field Studies. PHOTO BY AUTHOR.

TOP The Australian cassowary, a shy but occasionally dangerous bird that stands five feet tall. No one is certain what the casque—the bony crest on its head—is used for. PHOTO BY AUTHOR.

BOTTOM A spectacled flying fox. Highly gregarious, some colonies of these intelligent bats once contained hundreds of thousands of individuals, but today they are far less abundant. PHOTO © MERLIN TUTTLE, BAT CONSERVATION INTERNATIONAL.

TOP, LEFT A Queensland wood-nymph butterfly. PHOTO BY AUTHOR.

TOP, RIGHT A lemuroid ringtail possum. Though normally in the treetops, this youngster was found near the ground. It was very ill and eventually died, despite our best efforts to save it. Lemuroid ringtails rapidly disappear from forest fragments. PHOTO BY AUTHOR.

BOTTOM A coppery brushtail possum. This particular individual is unusually dark, especially its face. PHOTO BY GREG DIMIJIAN.

TOP The long-tailed pygmy possum is one of the prettiest rainforest animals—but only the size of a mouse. PHOTO BY MIKE TRENERRY.

BOTTOM As shown by this photo from an automatic camera, giant white-tailed rats attack bird nests. Highly opportunistic, they also eat seeds, fruits, insects, and other small animals. PHOTO BY AUTHOR.

TOP The dreaded stinging tree, with its heart-shaped leaves. A good hit from a stinger can hurt for months. PHOTO BY AUTHOR.

BOTTOM A Day-Glo rat. Covered in fluorescent dye, this bush rat left a bright trail that allowed us to track its nightly activities. PHOTO BY AUTHOR.

TOP KingKella dons traditional garb before attending a highland *sing-sing.* PHOTO BY AUTHOR.

BOTTOM Chris Buehler, Steve Comport, and Joey MacMillan enjoying the Millaa Millaa wet season. Steve is responding to my suggestion to "grin and bear it." PHOTO BY AUTHOR.

guerilla warfare

As August arrived, we faced the aftermath of the federal election. Bob Hawke's victory had ensured that the World Heritage nomination would proceed—or at least Hawke's government was committed to making it so.

Following the initial shock, many rural residents became even more determined to fight the nomination. They were digging in for a long, bitter battle. Anti–World Heritage signs began to spring up in front yards, bearing slogans like "Dump Hawke," "No To World Heritage," and "Go To Hell Greenies!"

The rancor of the local populace was fueled, of course, by conservative politicians. The ignominious collapse of Joh Bjelke-Petersen's election campaign had only increased his determination to fight World Heritage, despite the fact that he'd relinquished the premiership of Queensland during his election bid. His replacement, conservative Mike Ahern, seemed equally determined to fight the nomination. One of the first things Ahern did was to seek an interim injunction from Australia's High Court to block the nomination.

In north Queensland we began to see a lot of "panic clearing"—landowners bulldozing and logging their rainforest in the fear that the government would prevent them from doing so in the future. Several thousand acres were destroyed in a short time. It became nearly impossible to hire a private logging or bulldozing contractor; they were all overcommitted.

In an effort to slow the clearing—which was being fueled by a combination of misinformation, rising hysteria, and uncertainty over the boundaries of the World Heritage Area—Senator Graham Richardson, minister for the environment in the newly reelected Hawke government, released a map showing the "indicative boundaries" of the WHA. A lot of forest destruction was going on in areas that weren't even being considered for inclusion, and the hope was that the map would help reduce this.

A modest amount of private land had been incorporated in the WHA's proposed boundaries—there was no way to avoid it. In several cases, crucial corridors of privately owned forest linking major forest tracts were included. In others—such as that of our next-door neighbors in Millaa Millaa, who ran a small sawmill using timber they cut from their own land—it was a matter of protecting a very rare type of rainforest.

These were gut-wrenching decisions, of course, and I took no joy in the thought that people—neighbors, not just anonymous faces—could lose their livelihoods. The government allowed for written objections to its proposed boundaries, and set up a panel of eminent scientists to consider these. Once it was determined that a private block would be included, the government's goal was for the landowner to retain the land by signing a formal agreement not to clear or log their forest. Only if an agreement could not be hammered out, would "resumption"—government purchase of the land at reasonable market value—be considered.

Steve and Chris had done such a stellar job with the fieldwork that I began seriously considering something I'd long been wanting to do, that is, to visit New Guinea. Our project was drawing to an end, and I had no idea whether I'd be able to return Down Under. This might be my only opportunity.

Although funds were tight, I still had a little of the money I'd saved for graduate school, and the island of New Guinea lay just at Australia's northern doorstep. Flights from Cairns to Port Moresby, the capital of Papua New Guinea, were only three hundred dollars.

I've often traveled alone, but the thought of going solo to New Guinea barely entered my mind. Papua New Guinea was, and is, one of the most lawless places on earth. It has been estimated that in the past two decades, more people have died violently in PNG than in war-torn Beirut.

News of PNG atrocities rarely reaches the United States, but Australians are accustomed to horror stories. Only weeks earlier, Queensland newspapers had reported the murder of an Australian woman, a helicopter pilot and devout Christian who often helped the missionaries in PNG. She'd grown tired of the violence and had finally decided to leave. Although highly security conscious, she'd let down her guard momentarily while trying to sell her car. She went for a drive with a prospective buyer, and that was the last time

The island of New Guinea, divided into the Indonesian state of Irian Jaya, and the nation of Papua New Guinea. New Guinea is the largest tropical island in the world.

she was seen alive. Her body, gang-raped and brutally tortured, was found a few days later.

I spread the word that I wanted to visit PNG, and was looking for an adventurous soul to go with me. I also checked into the possibility of taking a pistol along, but was assured I'd never make it through PNG customs—and even if I did, that a pistol would incite more trouble than it prevented.

In late August I got a call from two young Americans, Jeff and Sue, who were studying marine biology at James Cook University. They didn't know a lot about New Guinea but thought the trip sounded great. We met to get acquainted, and began planning our adventure. I explained what I knew about the island.

New Guinea is the world's largest tropical island. Following a complex history in which segments of the island were colonized by the Dutch, English, Germans, and Australians, it emerged from World War II split into two: the western half, controlled by the Dutch, and the eastern half, controlled by the Australians. The Dutch abandoned western New Guinea when they lost the East Indies in 1949. Indonesia, the vast archipelago-nation that emerged from the iron fist of Dutch rule, soon laid claim to western New Guinea—backed by military force—which it renamed Irian Jaya.

The eastern half of New Guinea was managed in a more benign manner as an Australian protectorate, and was granted independence in 1975 as the nation of Papua New Guinea. Technically, PNG's government is based on a parliamentary system, just like Australia. In reality, however, it has a second, far more complex system of governance operating at another level entirely—the system of tribal one-talks.

Nearly every square inch of New Guinea is claimed by one tribe or another—each with its own particular language. With over seven hundred separate tribes, New Guinea is estimated to contain a third of the world's languages. This remarkable linguistic and cultural diversity is an artifact of the island's steep, dissected topography, which promotes a high degree of social isolation.

With so many languages, it was natural that with colonization came the evolution of a common tongue. New Guinea Pidgin is a strange, bastardized mix of English, German, Indonesian, and other languages and contains only 1,300 words in total. With so few words, even simple concepts can take a lot of verbiage to explain. For example, "haus buk bilong ol man/meri" means "library" (a house with books for men and women), while "numbawan pikinini bilong misis kwin" refers to Prince Charles (the first child of the queen). On an internal flight in PNG I once spent half an hour trying to decipher a hundred-word sign on the bulkhead that I eventually came to realize meant "Keep your seat-belt fastened while the plane is in the air."

Perhaps what is most remarkable about New Guinea is that the bulk of its native population—nearly a million people—remained undiscovered by Europeans until the 1930s. These were the myriad tribes of the New Guinea highlands, a vast mountain chain that runs roughly east to west, forming the spine of the island.

In 1931, a gold prospector named Mick Leahy first penetrated the dense jungles and hostile tribes to enter the montane interior of New Guinea. He was stunned to find large numbers of people living in the highland valleys. The highlanders had never before seen a white man, and were equally shocked at the encounter. Leahy both wrote about and filmed his discoveries, and was soon followed by other explorers and adventurers. Shortly thereafter, the British and Australians set up a series of administrative districts in the highlands run by "bush police"—hardy, field-toughened men who tried to keep order among the often-warring tribes.

It goes beyond the scope of this book to describe this fascinating era—

this collision of stone-age and twentieth-century societies. I simply must, however, mention one unique event—the evolution of the Cargo Cults.

The Cargo Cults were inspired by the newly formed colonial administration. The highlanders watched in fascination as the white men cleared a big flat area, built strange, square-shaped huts, and then spent much of their time passing pieces of paper back and forth. Then—to their great astonishment—a giant growling bird would drop down from the sky to disgorge crates of food and other cargo. It was clearly a miracle—a gift from the gods.

So the tribesmen made their own clearings and built their own little square huts, then spent their days endlessly passing bits of paper back and forth. The Cargo Cults had been born. It is a testament to unrequited faith that the cults persisted for decades throughout the highlands, despite the fact that the giant bird never once paid them a visit.

Notwithstanding the tribal cultures, it was the island's biology that most intrigued me. New Guinea has been linked to northern Australia throughout most of its geological history. The fauna and flora there are distinctly Down Under in character—marsupials and monotremes, birds of paradise, and relict Gondwanic[4] plants—with a few Asian representatives thrown in to make things interesting.

But New Guinea's rainforest biota is far more diverse than Australia's, which lost many of its species during the past few million years, as the continent's rainforests shrank into small, isolated pockets. Where north Queensland has three species of melomys, for example, New Guinea has ten or more; rather than two tree-kangaroos it has at least a dozen. There's nothing like such stunning diversity to capture the interest of a biologist.

An unexpected advantage of organizing a trip to a dangerous place like New Guinea was that nobody in Millaa Millaa suggested I was leaving because I couldn't handle all the strife. However, if I entertained any hopes that things might soon begin to calm down in Queensland, I was sadly mistaken. The

4. Gondwanic species arose from ancient lineages associated with the southern landmasses—principally Africa, Madagascar, South America, Australia, New Zealand, and Antarctica. These lands were once joined (along with the Indian subcontinent) into a supercontinent called Gondwana, which began breaking up about 100 million years ago to eventually form the continents we know today. Plant and animal groups that originally evolved in Gondwana are often confined to these southern regions. A classic example is the large, flightless ratite birds—the ostriches of Africa, the recently extinct elephant birds of Madagascar, the rheas of South America, the emus and cassowaries of Australasia, and the kiwis and recently extinct moas of New Zealand.

instigator of the latest troubles was the head rabble-rouser himself, Joh Bjelke-Petersen.

The final vestiges of civility seemed to collapse on a sunny afternoon in late August. I walked into the Millaa butcher shop to get a kilo of hamburger—"mince," the Aussies call it. The butcher and I often bantered as I made my purchases, but today his voice had a hard edge.

"Mrs. Davies and Mrs. Appington were just in here," he said (not their real names). "And they were telling everybody what a goddamned ratbag you were. A right scumbag, they reckoned. Show up here acting all friendly, then try to stab all the people who've helped you in the back. They reckoned you ought to get your ass kicked, proper."

At first I thought he had to be pulling my leg—he often did that. But it quickly sank in: the ladies had been truly furious. Even by recent standards, this was unprecedented anger and bitterness. If I'd been unpopular before, I'd just been branded Public Enemy No. 1.

What was going on *now*? I left the butcher's in a daze, sickened.

I soon found out what had everyone so riled up. Bjelke-Petersen had just visited north Queensland and in a wild rally in Cairns told two thousand supporters that World Heritage would "wipe out your future." "This is war," he proclaimed. "This is more dangerous and far-reaching than what happened in the Tasmanian Franklin Dam affair and it is a fight you northerners must win." He also incited timber workers to "harass federal Labor MPs" (members of parliament in Bob Hawke's party) and to "Ring them at all hours, call them at home." He closed by saying "We are at war . . . Put a fire under Richardson when he comes up here," a reference to Graham Richardson, Bob Hawke's minister for the environment, who would soon visit Ravenshoe, the logging town.

This infamous "war speech" became a turning point in the World Heritage controversy. If north Queensland's rural population had been upset before, now they were verging on violent. Bjelke-Petersen had virtually incited them to war, and the reverberations from his fiery speech would be felt long afterward.

The following day, Senator Graham Richardson was attacked in Ravenshoe. He was on his way to the town hall with a police escort when a huge angry mob blocked his path with a logging truck. Richardson wanted to stay in his car, but the mob rocked the car back and forth, threatening to roll it over. He finally got out and was made to climb onto the back of the truck to address the crowd. He spoke briefly—continually interrupted by jeers and boos—then tried to squeeze through the throng back to his car. On the way, he was shoved and jostled, and someone yanked his hair. Then a fist flashed out of the crowd and caught him in the side of the face. Only then, it seemed, did the Queensland police jump in and really try to protect him. Richardson was furious with the state police, whom he maintained just stood by and watched his attack.

In the following days Millaa Millaa seemed ready to burst apart. I was working with our crew again, trying to finish the trapping in one of our fragments before departing for New Guinea. Eight of us had loaded into the truck and were zipping along Middlebrook Road, a narrow, winding lane bordered by pastures and forest. We'd just come around a corner when our truck crashed down onto the pavement. Five people in back were thrown out of their seats. Sparks flew as I fought to keep us from spinning out of control. Amazingly, as we screeched to a stop one of our tires went rolling past us. We sat there, gasping, wondering what the hell had happened.

We got out of the truck and looked at the rear axle, which had dug a long groove in the pavement. The lug nuts that held the double tires to the axle had been sheared off. It suddenly hit us—someone had unscrewed all the lug nuts, leaving them just slightly attached so the tires would fly off as we were driving. This went *way* beyond threats and intimidation. Some son of a bitch had tried to kill us.

new guinea

As we boarded the Air Niugini flight to Port Moresby I felt vaguely uneasy, though also relieved at the idea of getting away from north Queensland for two weeks. As before, Steve and Chris were running things in my absence. On my return we'd go through the final stages of closing down the project.

The last week had been hectic, with fieldwork, getting the truck repaired, and preparing for this trip. I'd phoned the Millaa cop to report the lug-nut incident, but he didn't ask me to fill out a police report, and I didn't push it. Jenny had been nervous about our trip to New Guinea, and I'd assured her repeatedly that we'd be keeping our heads down.

Our trip nearly ended before it began. On takeoff our plane hit a flock of egrets that had flown up from the swamps near the Cairns Airport. We heard a bang from the left jet engine, then the pilots did a sharp U-turn and dropped straight back down to land. The engine's intake rotors were damaged, so our flight was delayed until another plane could be brought in the following day.

We'd expected to see jungle as we flew into Port Moresby, but the mountains surrounding the city were cloaked in monsoonal woodland—lots of dry grass and eucalypt trees. The customs official stamped our passports unceremoniously, and we were in PNG. We'd made only the vaguest plans for our trip; New Guinea isn't exactly set up for tourists.

We didn't plan to stay long in Moresby. A man we'd met on the plane suggested a cheap hotel, where we all crammed into a single room. The taxi ride

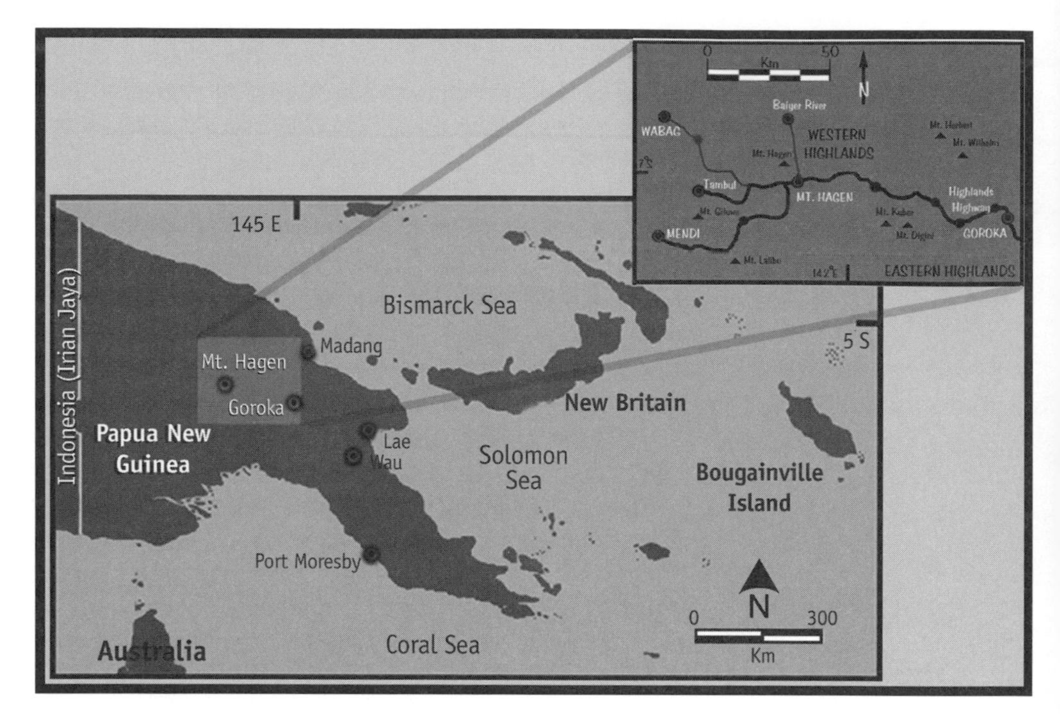

Route of our travels in Papua New Guinea—flying from Port Moresby to Wau, then traveling by buses to Lae and up into the remote highlands.

to the hotel had been an eye-opener, and the reality that we were three very white people in a land where virtually everyone was black began to sink in. Every house was surrounded by tall fences topped by razor wire. Dogs and guards stared out from within.

I have to give Jeff and Sue credit—they were nothing if not adventurous. On our first night we went dancing at a PNG disco, the band blasting out rock and country songs in a pidgin accent. As the disco closed the rowdy crowd flowed onto the street, and two drunken fistfights erupted. We took off at a jog and hailed a taxi.

We spent the following day exploring Moresby—a dingy city of a few hundred thousand ringing a ship-filled harbor. We chatted as best we could with people on the street, making an effort to converse in pidgin. We tried chewing betel nut, which is mildly intoxicating and turns your teeth a brilliant blood-red. In the afternoon we boarded a flight for Wau, high in the rainforest-clad mountains to the north.

The pilots in PNG—mostly Australians—certainly earned their salaries.

To land in the Wau valley the pilots had to plummet down in a series of tight spirals until they'd cleared the surrounding mountains, then land on a short grassy strip. The air was so dense with smoke it was like flying through a cloud.

The townspeople gawked at us—mouths literally hanging open—as we walked along the dusty, unpaved streets of Wau. We learned later that whites almost never walk in PNG—they drive. We stopped at a store and I decided to try the local tobacco, called *bruis*. This was a mistake; the tobacco is sold in big blocks heavily congealed in tar, which are broken up into smaller chunks and then rolled into fat cigars using a newspaper. I took a few puffs and nearly expired right then and there, gagging and gasping for air—a sight that the crowd following on our tails found enormously amusing.

We hitched a ride to the Wau Ecology Institute, which is located on a hillside at the edge of town. The institute has good labs and a dormitory, but only a couple of scientists; few had visited since the violence had grown out of control. Two other hardy travelers were staying in the dorm—an Australian helicopter pilot who was working for a mining company, and a tough-looking Israeli solider who was traveling alone.

The day after we arrived a chap named Steve Pruett-Jones returned from his field camp in the nearby mountains. I was pleased to catch up with him. Steve and I had gone to graduate school together, and he'd since taken a faculty job at the University of Chicago. For many years Steve and his wife, Melinda—also a grad student at Berkeley—worked together in PNG, studying the behavior of birds of paradise. Now, for the first time ever, he was here alone; it was just too dangerous for Melinda.

Steve told us about the latest drama in Wau. A few weeks earlier a man had raped a young girl, from another local tribe. The girl's one-talks (tribe members) had caught the man responsible, and chopped off each of his fingers, then his hand, and then his forearm, by which time he'd bled to death. They dumped his body on the main street of Wau. The man's one-talks reciprocated by sneaking up to the other village and burning it to the ground. Soon thereafter the PNG National Guard arrived, and a state of emergency was declared, ending the hostilities.

This incident vividly illustrated "payback," Steve explained, the deep tribal precept that no harmful act could go unpunished—quite like a Sicilian vendetta. Payback, however, has some uniquely New Guinean rules. First, because the tribes are so closely integrated, it isn't essential to catch the person who performed the wrongful act; if you can't find him, it is quite acceptable to kill one of his one-talks instead. Second, rounds of payback between tribes can stretch on for years—even decades. They can be settled only by a *sing-sing*, an elaborate festival in which the tribe that was last wronged receives a payment—usually pigs—in exchange for agreeing to end the conflict. It is ironic that the fundamentalist missionary groups in PNG have tried to ban the *sing-sings*, which they regard as pagan rituals. Apparently it is more important to save souls than lives.

We chatted with the Aussie helicopter pilot, and he offered to take us along while he checked out a landing pad in the mountains. The guy was a red-hot pilot—a veteran of Vietnam and of Australian cattle-mustering, which involves constant aerial acrobatics. As we climbed above Wau in his French Llama helicopter we could see huge fire scars climbing up the mountainsides. The natives were pyromaniacs, the pilot said, which explained why the valley was so smoky.

We blasted up to the mountaintops, skimming above stunted trees dripping with moss. The pilot offered to drop us off so we could look around, then return in an hour. He curled into a small clearing and hovered motionless, a foot above the ground, while we all leapt out.

From where we stood, the panorama of jagged mountains was stunning. We explored the forest along the edge of the clearing and thought about hiking down to a river we could hear below, but figured we might not make it back in time. For the first time in PNG we were seeing birds—they'd been virtually killed off in Wau and Moresby by the tribesmen, who are phenomenal hunters.

Suddenly we realized we were being watched. A handful of men had materialized only yards away, holding machetes and hand axes. The world's last documented cases of headhunting happened here, in the mountains above Wau, two decades earlier. Fortunately, the men seemed just as stunned as we were, and after a long stare they melted soundlessly into the forest.

Our pilot returned and took us on a roller-coaster tour of the mountains. He'd plummet down into valleys, then scream up steep mountainsides, the G-forces pinning us to our seats as he narrowly cleared the summits. He must've enjoyed our whooping and hollering, for he gave us one heck of a ride.

We left Wau two days later with a rough-talking Aussie expat we'd met in the town's "country club," which had only a handful of patrons—all white. The man drove like a maniac, swearing vehemently at the natives, who had to dive off the road to get out of his way. He had a machete and billy club under his seat—defense against rascals, he explained.

A beautiful ebony python from the New Guinea highlands, near Wau. PHOTO BY AUTHOR.

"Rascals" is an innocuous-sounding name for the murdering bands of thieves that terrorize the New Guinea highlands. The bands are comprised of young men carrying homemade shotguns and machetes. They're a lot like pirates, except that they attack in trucks and prey on cars and public buses—called PMVs (public motor vehicles). A gang of rascals will pull over a PMV, rob and often rape the passengers, and casually kill anyone who dares resist.

The rascal problem has dramatically worsened in recent years. There are several underlying causes. The population of PNG has exploded with the introduction of Western medicine, particularly owing to a sharp drop in infant mortality. This, along with the gradual breakdown of tribal society, has created many social problems. Disenfranchised youths have tended to accumulate in the highland towns, where they often find trouble. Compounding these problems is rampant alcohol consumption.

A few miles outside Wau we came upon two of the oddest contraptions I'd ever seen. The expat drove up to give us a close look, and we climbed onto one of them. They were great rusting heaps of metal—the size of battleships—with towering ramparts and conveyor belts. These were the famous floating dredges that had once scoured the Wau valley for gold—some of the richest gold deposits ever found. Unbelievably, the two dredges had been flown to Wau, piece by piece, in specially modified Junkers Tri-motors. For a two-year period just after World War II, more tonnage was flown into Wau than any other airport in the world.

The expat dropped us off in a town where we caught a PMV bound for Lae, on the northeastern coast. The PMV was designed to carry about two dozen passengers but held at least three dozen, and the driver kept stopping to pick up more. Everyone was carrying something; the woman beside me perched atop a sack of grain, while someone in the back had a squealing piglet.

As we dropped down into the lowlands the rainforest was broken by oil-palm plantations. Large forest tracts were being logged by Malaysian timber companies, who sold the logs to the Japanese—mostly for cheap plywood and disposable chopsticks.

We made Lae before nightfall, which was a relief given all the stories we'd been hearing about rascals. We were exhausted, and ended up crashing in an abandoned schoolhouse near the bus depot, sleeping fitfully on the floor as we sweated and jumped at every strange sound.

At around 1 A.M. I heard a muffled scream and lifted straight up out of my sleeping bag. There was a wild thrashing sound, then Sue shrieked again in panic. It was pitch-black but I could hear people moving all around us, then shouts in pidgin. Jeff yelled out frantically—they were dragging Sue off. I fumbled for my light but something slammed into the back of my head. I fell back, stunned. Jeff was grappling desperately with someone, then let loose a horrifying yell, and another. I managed to find my flashlight and through swimming stars saw Jeff clawing at his throat, which was gushing blood. Sue was being shoved out a broken window . . .

"Bill . . . Bill! Are you okay?" Jeff was shaking me in the dark.

"Oooohhh. . . . My god, Jeff . . . Just a bloody dream . . ." I was drenched in sweat, shaking. Jeff and Sue started laughing nervously. I lay back down and took some deep breaths.

It was one of the most vivid nightmares I'd ever had. And it wasn't to be the last. For months after PNG I had harrowing nightmares of being attacked, brutalized, killed. I considered myself adventurous, but I hardly ever stopped feeling scared. The danger was palpable—vibrating through the air, whistling in our ears. I was glimpsing what the American soldiers in Vietnam must have felt—unending tension; feeling conspicuous and vulnerable; and the very real prospect of sudden death . . .

One thing was becoming clear: I wasn't as adventurous as I'd thought—not even close.

missionaries, mercenaries, and misfits

The following day we caught a PMV to our main destination, the highlands of New Guinea's Central Cordillera. A few weeks before our trip, Sue had written to a friend of a friend in Mt. Hagen—the manager of a coffee plantation, whom she'd never met—and he'd kindly invited us to stay with him.

The Highlands Highway snakes its way up into sawtooth mountains that form the jagged spine of New Guinea. The climb was often torturously steep, the overcrowded bus sometimes groaning to a crawl. There was a surprising amount of traffic—PMVs, trucks piled high with people and goods, surly-looking characters on small motorbikes. Everyone stared at our white faces as though we'd just arrived from another planet.

We passed by dozens of small villages. The common theme was a cluster of grass huts in a central clearing, adjoined by fields of sweet potatoes, bananas, sago, and other crops. Well-fattened pigs—the true measure of a man's wealth—wandered about, occasionally joined by a tame cassowary. Each village had its own unique architecture: there were circular huts with flat roofs, rectangular huts with sloping roofs, octagonal huts with peaked roofs—every conceivable pattern. Often, sticking out like a sore thumb, was a Christian church, also constructed of grass but bigger than the other huts and with a rigid, European geometry.

The pall of smoke was everywhere; like Wau, the valley bottoms were grassy and maintained by annual burning, massive fire scars climbing the steep mountainsides. People were noticeably stockier here in the highlands.

In one village we saw a rugby game contested by barefoot youths, and were impressed by their vigorous play.

While the scenery was intriguing, a recurring shock was seeing live wildlife being sold for food. A common method of advertisement was simply a wooden spear stuck into the ground along the roadside. Lashed to the spear would be a tree-kangaroo or cuscus (a rainforest possum) that had been bashed half-senseless, twitching in agony. While this seemed unthinkably brutal at first—why not put the poor beast out of its misery, then simply sell the meat?—we soon realized that there was no refrigeration.

We were also surprised to see marijuana growing along the roadside. Jeff and I debated this for a while—he insisting it was pot while I couldn't quite believe it—until we passed right by an eight-foot plant. We found out later that, while technically illegal, pot was sold quite openly in the village markets. It is called *spark bruis* in pidgin—tobacco with a spark.

We arrived in Mt. Hagen at the end of another long day and stood in a queue to use a pay phone. As in Wau, a gawking crowd trailed behind us—a most disconcerting feeling. The plantation manager, Robert, soon arrived in a shiny new four-wheel-drive, giving us a cheery g'day as he simultaneously scowled at the natives.

Robert's house, a few miles outside Mt. Hagen, was palatial. He had a swimming pool, satellite TV, and five-thousand-dollar stereo, he explained breezily. He also had a cook and a houseboy—"dogsbody," he called him—and a fierce-looking warrior with bow and arrows who prowled around his razor-wired yard to repel intruders. On his forearm Robert bore a huge scar where a worker he'd fired years before had attacked him with a machete. He explained with a wink that he'd arranged for the man to take a little helicopter ride—his first and last—and we could tell he was utterly serious.

Robert offered to take us on a tour of the plantation. We tore over bumpy tracks as native workers dove out of the way, Robert swearing incessantly out the window. He stopped to talk to one of his foremen, vigorously abused the man in pidgin, then turned to us with a disgusted look, his message clear: *See the idiots I have to work with?* He seemed to be trying very hard to impress us.

That evening Robert took us to dinner at the local country club—like Wau, the patrons all white expats. We drank heavily and listened to blaring music, and Robert complained in an increasingly slurring voice about how badly PNG was deteriorating. Halfway through dinner he decided he didn't like a particular busboy, who apparently wasn't being obsequious enough.

He grabbed him by his white jacket and rammed him up against a wall, staring evilly into his eyes and swearing long oaths in pidgin.

I snapped. I tore into Robert, heaping him with abuse. Sue tried to intercede, but it was too late. I don't remember what I said, but he seemed stunned by my outburst. What was I upset about? He was just giving one of the blacks some hell. We left the club in an uproar, Robert saying over and over that he was going to make me respect him.

Sue was visibly annoyed with me, and we rode back to Robert's house in tense silence. When we arrived I decided I'd had my fill of this jerk, and starting packing my things. Sue and Jeff begged me to be reasonable—where could we go in the middle of the night in the PNG highlands? Then Robert grabbed my arm and dragged me into the living room—I *had* to hear something. He pressed a button on his stereo, and a crooning voice suddenly filled the room at deafening volume.

Robert's triumphant expression told us he'd been vindicated—that a great personal truth was being revealed. Then his expression changed. He became oddly somber, then shuddered, suddenly hiding his face as great sobs racked his body. We stared at him in disbelief. Here we were in the PNG highlands, in a palatial house in the midst of one of the most primitive wildernesses on earth. Our host, a rednecked plantation-master, was crying himself to sleep while Frank Sinatra sang "I Did It My Way." "Bizarre" doesn't begin to describe it.

Though sporting hangovers, we packed up early the next morning and returned to Mt. Hagen. Sue wasn't happy to leave our safe and opulent surroundings but Jeff and I were determined. Someone had told us that there were only three types of expats in PNG these days—missionaries, mercenaries, and misfits—and Robert clearly qualified for two of these.

We strolled around Mt. Hagen, a frontier town of a few thousand, and were struck again and again by the anachronisms. Women wore combinations of traditional clothing and Western garb, some bent forward under heavy loads in string bags called billums, the straps supported by their foreheads. One man wore "arse-grass"—palm leaves that barely covered his groin and rear—along with tennis shoes. An old woman selling soft drinks had had the outer joints of three fingers cut off. We later learned this meant she'd lost three babies; infant mortality had been very high a generation earlier, and this was how women expressed their grief and ensured that each child was remembered.

We arranged to take a PMV northward to Baiyer River, a settlement with a zoo of sorts. We'd been warned that the Baiyer River area was a "Declared War Zone," which meant open fighting had broken out between two tribes. If anything this was an incentive for us to go—these were traditional warriors, not rascals—and we were keen to see them.

A few miles from Baiyer River our PMV suddenly stopped. Ahead of us a queue of buses and trucks had formed, just like at a railroad crossing. We stuck our heads out the windows and soon glimpsed men in arse-grass running around a few hundred yards ahead. After several minutes, the line of vehicles began to move: the warring tribes had called time-out so the traffic could pass. As we drove by, a few warriors dashed forward to improve their positions. One heaved his spear across the road just as the last PMV drove by.

We shook our heads in amazement. The highland tribes of New Guinea might be primitive by many standards, but their wars—complete with breaks in the action to accommodate traffic—were remarkably civilized. We'd heard that the traditional conflicts often involved more bluster and screaming than actual killing. If someone got a spear in the foot, the battle was often called off—the injured side swearing to even the score at some future, unspecified date.

The zoo at Baiyer River was dilapidated, but it gave us a chance to see some of the spectacular birds of paradise. The males of each species have wonderfully garish plumage and equally wild mating displays; in some species the male literally hangs upside-down from a branch while madly flapping its wings. There were other native species too—hornbills, cuscuses, tree-kangaroos. I followed a half-tame cassowary around, snapping photos, when it suddenly pivoted and struck a threat display—one foot aloft with its daggerlike central toe poised to strike. I froze, and it turned and walked away.

On the return to Mt. Hagen we struck up a conversation of sorts with our PMV driver, who spoke a little English, while we attempted pidgin. His name was KingKella, and something about him suggested he was a *bigpela*—somebody important. He invited us to visit his village and was delighted when we quickly accepted his offer.

KingKella was from the Togoba tribe, which occupies a series of villages at the base of a great massif about ten miles outside Mt. Hagen. Missionaries had set up a grade school near the main village, where some of the children learned basic subjects and rudimentary English. The Togobas lived in

This happy gang of Togoba kids followed us everywhere, and laughed uproariously when I clumsily tried to hand-wash my clothes in the river. PHOTO BY AUTHOR.

square-shaped huts, but KingKella's was different—a hybrid between the traditional design and something larger and more modern. Instead of the usual grass, his walls were made of woven mats, and the hut was subdivided into several rooms, one of which we had to ourselves.

We became instant celebrities among the Togobas, who overwhelmed us with their kindness and sincerity. In fact, they eventually adopted us as honorary one-talks—a fact announced by a tribal elder who informed us we were now "PNG pelas tru." This was just the kind of experience we'd been hoping for.

The tribe's cohesiveness was remarkable. We befriended a young man named Dokta, who knew a bit of English. One day I pointed to a teenage girl and asked who she was.

"Sista," he replied.

That's nice. "And that one there?" I said, pointing to another girl.

"Me sista, too. And she, and she, sistas too," he said, pointing to others.

They certainly had big families here. Dokta continued pointing out sisters and brothers for the next ten minutes. After twenty or so I began to see a pattern.

"But Dokta, these sistas and brothas no have same mama as you?"

"Nooo," he laughed. "Not that kind sistas and brothas. These one-talks," he said, as though such a distinction were trivial.

The Togobas bathed and washed their clothes in a river at the bottom of a five-hundred-foot gorge. They raced up and down the gorge like mountain goats, and we panted in the thin mountain air trying to keep up with them.

On the second day they took us to wash our clothes, with a big mob of kids following along to watch. I'd never washed clothes in a river before, and my technique must have been pretty awful, because everyone seemed to find it hysterically funny. Finally, with theatrical flair, one of the older kids picked through my bag of filthy clothes, constantly grimacing and wrinkling his nose. He pulled out a truly disgusting pair of underwear, which he proceeded to hold up front and back so everyone could get a really good look. Then his hands shot into the water in a big froth of soapy bubbles, and in twenty seconds my underwear emerged sparkling white. He proudly displayed these and took a deep bow as everyone applauded.

The affection everyone showed us, and each other, was stunning. Everyone held hands. It took me a while to get used to holding Dokta's hand, but he seemed so crestfallen when I hesitated that I soon gave in. We never heard a word spoken in anger—never saw a child spanked or yelled at. It was like being part of a huge, happy family—a feeling of remarkable security. After our nerve-wracking journey in PNG, it felt like coming home.

the togobas

Each day with the Togobas brought new adventures. We hiked to a half dozen Togoba villages, led by Dokta and his friends and trailed by curious, laughing kids. We were received like visiting dignitaries. People rushed out to see us, shaking our hands or just staring guilelessly, openmouthed. We were also introduced to the *bigpelas*—elderly men—who presided over each little village.

Compared to the town of Mt. Hagen, just ten miles away, the villages seemed much more traditional—less affected by the culture clash of the twentieth century. There was a rough division of labor. Women and children worked the fields, while the men hunted, tended their huts and pigs, concerned themselves with village politics, and, occasionally, made war.

All of the young boys were skilled hunters and bushmen, and all could start a fire by rapidly pulling a strip of pliable tree bark back and forth through a split stick, creating enough friction to ignite some dried moss. Many carried a machete or hand-ax, honed to razor sharpness. It seemed that virtually any animal was fair game, an impression reinforced by the remarkable paucity of wildlife near the villages, even of small lizards and birds.

We remarked to Dokta that we'd like to see more animals, and he immediately said he could take us to a place where this was possible. With no preparation other than grabbing our binoculars and canteens, we took off, trailed by the usual gang of kids. We assumed we'd be hiking for a quarter of an hour or so. Dokta led us down into the steep gorge and straight up the towering, rainforest-clad massif on the other side.

After an hour of rigorous climbing, we asked Dokta how much further we had to go. "Not far," he said. We continued on for another hour, the sweat pouring out of us. Most of the kids had turned for home. Again we asked if we were getting close. "Not far," he repeated. After yet another hour the forest became stunted and dripped with moss, the terrain almost vertical in places. We dragged our way up muddy, slippery slopes, rain pelting down. Verging on exhaustion and getting extremely cranky, we repeated our question. "Not far," Dokta said yet again, looking slightly guilty this time. Jeff and I had just sworn an oath that if Dokta said "not far" one more time, we were going to throw him off the mountain, when we finally broke out of the forest.

We had cleared the summit. Before us was a small clearing—maybe an acre in size—with sweet potatoes and plantain bananas and a small grass hut. Famished, we cooked green bananas and sweet potatoes and gorged ourselves, and by then night had fallen. The half dozen of us who'd made it to the summit crammed into the little hut and fell asleep. We were warmed by a smoky fire that, by morning, made my throat feel like I'd chain-smoked five packs of cigarettes.

Dokta had arisen before us and for breakfast he'd caught us four little songbirds—no bigger than sparrows—which he'd found in a nest some distance away. The birds were still alive, suspended by a string tied to their legs that hung from Dokta's belt. Every time he moved they flapped and chirped helplessly. It was pretty hard for a bunch of animal lovers like us to watch, and Sue just couldn't take it. She demanded that Dokta let the little birds go. I reminded her that fledglings like these probably couldn't survive on their own, but she was adamant. After much hemming and hawing, Dokta finally complied, and the birds flitted off into the forest. We ended up having cooked bananas for breakfast, which suited us much better than little tweety birds anyway.

Descending the mountain was far easier than climbing it—and this time we knew what to expect. We'd had quite an adventure, though we'd seen no wildlife aside from fleeting honeyeaters and fruit-doves. In fact, the four little songbirds that Dokta had caught were the only animals we got a good look at.

The next day I made a tactical error. Accompanied by the usual mob of kids, we were relaxing by the river when Jeff turned to me with an excited expression. Why not take out one of my contact lenses and show them? Surely they'd never seen one before. That seemed like a fun idea, so I motioned for everyone to gather around me. The kids' eyes bulged as I slowly reached up and touched my eyeball. Then I grabbed the soft lens and dramatically peeled it away, holding it up for everyone to see. Their astonished faces looked like they'd just seen the devil himself. If I'd yelled "Boo!" I think they would have stampeded off in panic.

Word spread like wildfire, and from that point on I had to repeat my trick everywhere we went. We'd walk by a field or a hut and a dozen people would come galloping over, all vigorously pointing to their eyeballs. Then, surrounded by a wall of panting faces, I'd theatrically extract my lens, everyone gasping in wonder. What they didn't appreciate, of course, was the awful time I had getting the lens back in my eye without a mirror. My eyes were soon red and sore from the abuse.

We were approaching a Togoba village with KingKella when our nostrils caught an unmistakable smell: burning marijuana. The source turned out to be an old woman—seemingly in her seventies but probably much younger—who was puffing on a finger-sized joint as she plodded along under a load of firewood. Jeff's eyes lit up, and he asked KingKella if he knew where to get some more of the *spark bruis.* KingKella mumbled something indecipherable and I thought nothing more about it.

That night, in KingKella's hut, we were sitting around a circle with a group of village elders who'd come to see us. Dokta, who spoke about a hundred words of English, was trying very hard to interpret for us, and we chipped in with our very limited pidgin. The elders seemed puzzled about many things—What was happening in PNG? What was the outside world like?—and clearly wanted to hear our opinions. Then KingKella abruptly entered and tossed a football-sized object at our feet. It was a huge bag of marijuana—dried and ready to smoke.

KingKella tore off a page of newspaper, grabbed a huge fistful of pot, and rolled a joint that was easily a foot long and an inch in diameter. Jeff and I nearly keeled over. Only in Cheech and Chong movies had we ever envisioned a joint that size. KingKella handed it to Jeff and lit it. There was a great puff of smoke, Jeff's eyes bulging as he inhaled. He held it for about two seconds before exploding in a fit of coughing.

The joint was handed to me. Though I had abstained from pot in Queensland, there hardly seemed any risk here, so I took a hit too. Like Jeff, I gagged, surprised at the great volume of smoke that instantly filled my lungs.

KingKella was greatly pleased with himself. He and the elders waived off the joint, but kept encouraging us to smoke more. Within a few minutes the air was dense with pot smoke. Jeff and I were reeling, our perceptions stretched and distorted. Even normal things seem strange when you're stoned—and nothing was the least bit normal here.

The questions from the elders kept coming. After a great deal of confusion and miscommunication, we finally began to understand what was worrying them so much. It was the Baptist preacher who ran the church in the main Togoba village. He kept telling them that the world would soon be coming to an end. But he would never tell them *when*. Please, they asked—could we *please* tell them—when was the world going to end?

No episode of *The Twilight Zone* was ever so bizarre. These poor bastards! What could we say? They were utterly sincere—and desperate in the face of their dogmatic preacher and his all-supreme God. Only a generation earlier, these elders had had the confidence and wisdom to solve their own crises—they were the ones to whom others turned for counsel. But today, their world had turned upside-down. They were as confused and impotent as children. And they were turning to us for help.

Jeff and I gazed at each other, shaking our heads. We finally told them that not everyone believed the world was coming to an end—that the preacher might well be mistaken. This was a clear relief for the elders, though they still seemed appallingly confused. With our great difficulties in communication, it was all we could really do.

Aside from wildlife, the thing we'd most been hoping to see was a traditional highland *sing-sing*. As the principal means of settling disputes and ending potentially endless rounds of violent payback, the *sing-sing* has long played a vital role in the dynamics of New Guinea societies. The general idea is that the tribe that struck the last blow must compensate its enemy to end the violence, usually through the payment of pigs. *Sing-sings* may be sponsored for other reasons as well, such as to pay off the price of a bride.

The traditional *sing-sing* is a wild affair in which garishly decorated participants dance to drums, shaking the earth with their stomping and chanting. Each tribe has developed a unique style of dress, involving elaborate headdresses festooned with bright feathers, and bodies glistening with oils, paints, and pig grease.

KingKella delighted us by finding out about a real, traditional *sing-sing* that would take place the following day, about twenty miles from Mt. Hagen, near the town of Tambul. He organized a two-ton truck and pickup, and we and about thirty Togoba men piled into them.

We arrived at a remarkable scene. A dozen elderly and middle-aged men comprised the core of the celebration. They were adorned in traditional finery: faces fierce with red and yellow stripes, noses pierced by long sticks, their privates covered only by arse-grass. One muscular fellow wore an elaborate headdress and was holding center stage. This was the chief who'd sponsored the *sing-sing*, and he was shouting nonstop in pidgin. Tethered near him was a monstrously fat pig, the biggest I'd seen in the highlands. We later learned from KingKella what he was yelling so emphatically, over and over in many different ways:

> Look at this very fine pig I am giving away today! This is a wonderful pig—the best of all my pigs. This shows what a powerful and noble chief I am, that I would give away such a pig. I remember the last *sing-sing* we had with this same tribe, years ago. I received only two little pigs, which was a great disgrace. I am giving away this enormous pig, and there can be no doubt of what a wealthy man I am. Only a great and powerful chief would give away such a pig.

In addition to the main participants, there was a crowd of around two hundred onlookers, all New Guineans. Some had painted faces and pierced noses and wore the colorful woven hats of the highlands, while many others—especially the younger ones—were in Western garb.

Many young men had been drinking, and there was considerable tension in the air—a feeling reinforced by the fact that every male carried a razor-sharp machete or ax. The thirty-odd Togobas who'd come with us were very much on guard, clustered together on the periphery of the crowd. I was mesmerized by the proceedings and darted around snapping photos.

Trouble struck before I had any idea what was happening. I was taking photos of the crowd when a stocky little man in black facepaint went berserk, screaming at me in pidgin. He was wildly drunk and furious at me for taking photos without paying or asking permission. In an instant I was surrounded by a hostile mob of his one-talks, all jostling and yelling.

The Togobas heard the ruckus and came thundering over, weapons poised for battle. They shoved their way into the mob and one of them swung his ax right at the face of the screaming man who'd started the whole thing, stopping just an inch from his face. I was shoved to the back as the two groups faced off, toe-to-toe, weapons poised to strike.

My life passed before my eyes. I really thought, *Well, this is it. It's been an*

Me with the New Guinea highlands chap who nearly started a war. After almost getting me butchered, he decided to introduce me to his family. If you look carefully you can tell I've just had the living wits scared out of me. PHOTO BY A BYSTANDER.

interesting life, and now it's time to leave this world. I had this startlingly clear vision of all hell breaking loose and me trying to defend myself against swinging machetes and axes with my 35 mm camera.

The tension was incredible. Everyone in the *sing-sing* was watching. The two gangs seemed ready to murder each other; all that was needed was some tiny provocation—an inadvertent sneeze—to start the axes flying.

I figured I had to do something. Adopting a very contrite and humble expression, I waved at the stocky little fellow who'd started the conflagration. "Hello, hello—I'm truly, very sorry to have offended you," I uttered in shaky English and pidgin. "I had no idea I was doing something wrong. Please accept my apologies!"

For a long moment he stared at me. Then he simply shrugged, as though it were no big deal. As quickly as it'd started, the weapons dropped and the tension drained away. Then—unbelievably—he wobbled over and took my arm, saying he wanted to introduce me to his family!

I shot photos of him and his one-talks, all smiling and laughing. Knowing I wanted to remember this guy forever, I had a bystander snap a shot of us together, arm-in-arm like great old mates (this photo adorns my office today). Then he led me around the crowd to be sure that I got every shot I wanted.

The *sing-sing* and its heart-stopping standoff—brinkmanship, New Guinea style—stand out as my most vivid memory of our trip. And of one thing I have no doubt: the Togobas had been ready to fight a deadly battle on our behalf. Though we'd stayed with them but a week, they considered us real one-talks—*PNG pelas tru*.

Our two weeks in New Guinea were over, but we simply weren't ready to leave. We drove into Mt. Hagen and rebooked our flights to give us another five days. I tried to phone Jenny so she wouldn't be worried, but a fire had just destroyed the Telecom exchange in Port Moresby, making international calls impossible.

In our remaining time we'd planned a quick trip to Madang, on the northern coast several hours away. But our PMV was halted by a police roadblock just a few miles outside Mt. Hagen. No traffic to Madang was being allowed, we were told, because a tribe near Madang was killing anyone from Mt. Hagen. Someone from Mt. Hagen had accidentally run over and killed a child there, and her one-talks were getting payback.

We decided instead to visit Mendi, a small town in the southern highlands. This turned out to be a mistake. The PMV trip was interminable and overcrowded, and a massive man—much bigger than me—terrorized everyone on the bus. He yelled and bullied, slapped a few people around, and generally made the trip miserable. Mendi was a disappointment too, so we left after a single night and returned to our friends, the Togobas.

Our good-byes to the Togobas were remarkably sorrowful. Dokta was tearful and gave me his knitted highlands hat as a gift. We'd taken many photos and would mail them copies—mementos of our brief but unforgettable time together.

As planned, our final day was spent at the National Independence Day celebrations in Port Moresby. This was an extraordinary affair, with at least fifty tribes represented. The dancers stomped to beating drums, wearing spectacular decorations and headdresses. Many were clad in traditional arse-grass, the women bare-breasted. Others—the ones already converted to Christianity—hid their bodies beneath T-shirts and shorts.

On the flight back to Cairns it struck me that, during the past three weeks, I'd barely thought about Millaa Millaa and the World Heritage battle. Now it was time to face it all again.

return to oz

As we entered the airport lobby Jenny hugged me fiercely. She'd been terribly anxious about our trip, and when we failed to return on schedule she and Shirley were frantic—certain we'd fallen victim to murdering rascals. I'd finally managed to reach her that morning, from Port Moresby, and she'd cried with relief.

Driving into Millaa Millaa, I was shocked at how much things had changed during my brief absence. Anti–World Heritage signs had sprung up everywhere—on roadsides, street signs, telephone poles. Some houses were festooned with a dozen or more angry placards. The town looked like it was under siege.

Arriving home, I saw that Chris had the fieldwork well under control, assisted by a handful of volunteers. After seven months, Steve had recently left the project to become an intern for the School for Field Studies, where he'd be handling logistics and helping the students with their research. I'd assured the new director of SFS, Mark Noble, that Steve was fantastically qualified for the position.

There was much to be done over the next three weeks. We had to finish up the trapping and spotlighting, upload all the data, sell most of the research gear and other stuff we'd accumulated, clean the rented house from top to bottom, then pack up and leave.

Chris and I went about the melancholy task of our final trapping sessions. At our Zillie Falls fragment, we stared for a long time at the magnificent waterfall, mesmerized by the pounding torrent and wondering, sadly, if we'd ever see it again. Our very final session, on Mt. Fisher, nearly ended badly. Recent rainstorms had transformed the steep slopes into a slippery morass.

We were lugging traps along a ridgeline when I lost my footing and careened straight downhill into a big patch of stinging trees. When I glimpsed the stingers I instinctively curled into fetal position and—like a giant bowling ball—flattened several before finally skidding to a halt. Chris was horrified—certain I'd been severely injured—a perception enhanced by my less-than-stoic swearing and groaning. To my great good fortune, however, I'd been wearing a full-length rain parka that protected most of my body. I sustained painful hits on my ear and wrist, but otherwise I was intact. My ear burned so fiercely that I even considered a visit to the Millaa Millaa ambulance center, but—recalling the sadistic treatment poor Patrick had received—I decided to give it a miss.

In addition to all the logistical chores and last-minute fieldwork, there was another important thing I had to do—have a serious talk with Jenny. I'd been agonizing over what to do about this, and had finally reached a decision in New Guinea. If she were willing, we'd make every effort to stay together, even though I had to leave Australia to finish my thesis. We talked it over that night, and she agreed, teary-eyed. Yes, we'd stay together, and in a couple of months she'd come to visit me in the United States.

I was having a shower at Jenny's place when I heard furious yelling outside. One of the voices was Jenny's. I'd just managed to throw on my pants when Jenny came storming in, boiling mad. Her landlord had just kicked her out of the house. The SOB had set up two anti–World Heritage signs right in her front yard—a blatant provocation—and she'd promptly gone out and knocked them over. They'd had a big confrontation. Jenny and Kasie were out, he said, and he didn't care where the hell they went.

I was stunned—what a lowlife stunt! I felt like punching the guy's lights out. Of all the nasty things that had transpired in recent months, booting a young mother and her baby out of their home was the absolute limit.

Bloody Millaa Millaa! I loved it here—the rainforest, the wildlife, and, despite everything, many of the people too. But in a way it was a great relief to be leaving. And if things were hectic before, they'd just become frenetic. On top of everything else, we now had two houses to move out of.

It was a sprint. Chris and I worked sixteen-hour days, organizing and packing and uploading data, often working past midnight. We'd captured over six thousand mammals and, for each, we had to input over twenty items of information.

Over the past year and a half we'd accumulated an amazing pile of stuff—traps, field clothes, house stuff, arcane equipment. Anything I couldn't carry I had to sell. Fortunately, a local biologist wanted the research gear, especially the hand-made traps that had worked so well. The pickup, which had endured incredibly rough treatment, went back to the used-car dealer who'd originally sold it to me. I got nearly half of what I'd paid for it—a great deal as far as I was concerned.

Tully was a tough issue—what to do about him? I direly wanted to take him with me, but he would have to stay in quarantine for months, and that seemed plainly cruel. Fortunately, Shirley came to my rescue. She said Tully could live on the Taylor farm as long as needed—though I hoped to come back one day and take him off their hands. I would miss him enormously.

In the midst of trying to finish a hundred things, the impact of leaving finally hit me. I realized I was feeling overwhelmingly guilty. I'd arrived here *so determined* to stay on everyone's good side—and ended up widely despised and vilified. And, while I was simply leaving everything behind, the true-blue friends who'd staunchly supported me—especially the Taylors and Geoff Downey—would continue to bear the wrath I'd inspired. Sure, they'd made their own decisions to support World Heritage, but I was certain they would never have gone so far—or provoked such profound bitterness and ostracism—were it not for me. These thoughts gnawed at me constantly.

I could think of only one thing to do: face the people of Millaa Millaa. Perhaps we might achieve some understanding—some closure—on this whole mess. An idea began forming: I could give a slide show about our research findings. Then we could open things up and have a frank discussion about World Heritage—paying special attention to all the rumors and misinformation that had been flying around in recent months. The notion had some appeal: part research seminar and part town meeting.

Millaa Millaa has a town hall of sorts, a cavernous old Quonset hut. I arranged to rent it, then made fliers advertising the talk, which I posted all over the town. I'd scheduled the meeting for my second-to-last night in Millaa. The only question was whether anyone would show up.

As the fateful night approached I was horribly anxious. I'd heard through the grapevine that people were indeed coming. There'd been talk about boycotting it but they'd decided to attend, the discussion about World Heritage an obvious lure. But I had no idea how they would behave. Would someone stand up and castigate me, the whole thing spinning out of control? Might they become angry, abusive—even violent?

I'd nearly had a big setback the day of the talk. I'd been counting on the SFS gang to show up that evening—moral support I badly needed. But Mark Noble said they'd be visiting the coast and wouldn't be back in time. I'd put tremendous pressure on him to make it—promising that the real-life drama would be an invaluable lesson for the students—and, to my great relief, he finally agreed.

The meeting was scheduled for 8 P.M., and by the appointed hour the hall was crammed to capacity. At least two hundred people were there, a tremendous number given the area's sparse population. Some had come from Ravenshoe, the nearby logging town that was ground zero for anti–World Heritage activity. The crowd was solemn—tense—except for a few drunk yobbos who remained outside, firing angry glances at me when I'd stepped out for a smoke.

As I stood up to talk—thanking everyone for coming—I was sorely tempted to justify my past actions. I wanted to say how much I *liked* everyone here, how I'd never wanted to see anyone hurt. Nothing I'd done was ever *personal*—it was all about conserving the rainforest. But I realized how ridiculous that would sound. When you threaten someone's livelihood it *is* personal—regardless of what your motives might be.

For the first few minutes I was nervous, but once the lights went out and the slides appeared on the screen, I relaxed. I described the history of deforestation and fragmentation in the region, then explained how the rainforest's complex ecology begins to unravel in a forest island. Fragments are battered by winds, trees are destroyed and damaged, vines and exotic plants proliferate. Then the animals begin disappearing—quolls, cassowaries, musky rat-kangaroos, and some possums and antechinuses. Larger native predators like pythons, quolls, and rufous owls are replaced by small, opportunistic species—white-tailed rats and boobook owls. Rare, rainforest-dependent marsupials decline. The forest is simpler, poorer, impoverished—different.

The audience was into it, I could tell. No one had been the least bit disruptive. I finished by stressing the dire need to save the remaining fragments, especially those on rich basaltic soils. And to begin linking the fragments to each other and to surviving forest tracts with corridors of vegetation. Such corridors were invaluable, our research had shown, in helping to reduce the destructive erosion of animal diversity.

By the time the lights came back on I was feeling pretty good—though still anxious about the freewheeling exchange to come. I threw open the discussion by saying we'd all heard a lot of things about World Heritage—what it would mean, how it would affect the loggers, the farmers, the everyday person on the street. This was a chance to talk about it, to ask tough questions and get straight answers. John Winter, a prominent Queensland biologist and World Heritage advocate, had shown up, and I said that he and I and perhaps others here would try to respond.

Over the next hour and a half we covered every conceivable question. Who would be directly affected? (rainforest loggers and a handful of property owners within the World Heritage boundaries). Would communists be involved? (no!—just the federal government, and local communities would have a major voice in the region's management). How much money would be spent on displaced timber workers? (about fifty million dollars). Would all the timber mills close? (no—they'd be modified at government expense to process plantation trees, like Caribbean pine). Would the government try to take over farms outside the World Heritage area? (absolutely not).

After a while there seemed to be a collective sigh from the audience. The anxious edge was dissipating. We'd talked over everyone's worries and fears and tried to give no-nonsense answers. The discussion started veering off onto other topics—how to get rid of feral pigs, what to do about pot-growers. There were even a few laughs.

It was time to bring things to a close—we'd started at 8 P.M. and it was now 10:30. I thanked everyone again for coming and for putting up with me and my crew over the past year and a half—and for kindly letting us work on their land. I wished them all the very best for the future.

As the audience filed out, Chris overheard an elderly fellow—a retired Millaa logger—saying that maybe World Heritage wouldn't be such a great disaster after all. It was a small but significant victory.

Exhausted, we spent our last night in Millaa Millaa on a mattress on Jenny's floor, Kasie sleeping between us. We'd moved all her furniture to her parents' farmhouse, where she and Kasie would be living.

The next morning we drove out to say goodbye to Geoff Downey, then to see Steve, then to Millaa Millaa—Christine, Des and Helen at the takeaway, our young neighbor Nicole and her family. I tried to absorb the essence of the place, to burn it into my memory. Aromatic flowers, honeyeaters, the rainforest-clad mountains. There was something about this area—something almost indefinable—that seemed so ancient, as though at any moment a dinosaur might appear in the mist-shrouded forest, munching on a tree fern.

Jenny drove me down to the airport at Cairns. We said our good-byes, promising to call often. I took a long final look around, then boarded the 747, pressing my face against the window to hide my tears. The plane rose up from the mangroves and curved out over the Coral Sea. The last thing I saw was the aquamarine arcs of the barrier reefs, their colors so brilliant they seemed lit from within.

POSTSCRIPT

Today, the forty-two volunteers who made the Rainforest Fragmentation Project such a phenomenal success have scattered to the ends of the earth. I still get little snippets of news now and then, though the only ones I hear from regularly are Chris and Steve.

When the project ended, Chris returned to Switzerland, where she works as a florist. Once a year she leaves on a great sojourn to some remote area; twice she's returned to Australia. Steve—the once-frustrated machinist—finished both B.Sc. and M.Sc. degrees in zoology at James Cook University, and now works for the CSIRO Tropical Forest Research Centre in Atherton. John and I fell out of touch; the last I heard (a decade ago) he was working on a fishing trawler in Alaska.

Jenny did indeed come to visit me, and I took a break from thesis-writing to take her on a long tour of the western United States, before she returned to Australia. Our relationship eventually faded, however, the victim of too much distance and time, though we stayed in touch for many years. Shirley left Millaa Millaa after her husband died several years ago, but later moved back there.

I did return to Australia, and eventually wedded another Aussie, Susan, to whom I am very happily married. Tully lived with us for nearly a decade and now, at the ripe old age of thirteen (that's ninety-one in dog years), has been kindly adopted by my in-laws in Brisbane. After finishing my doctorate I was first hired as the director of SFS in Queensland, then became the lead scientist of the agency (the Wet Tropics Management Authority) that was created to manage the World Heritage rainforests of north Queensland. Aila Keto—the queen of Australian conservationists—was on my interview panel, and I dare say that was an advantage. A few years later I accepted a research fellowship at the CSIRO Tropical Forest Research Centre, before moving to Brazil with the Smithsonian Institution—where I've even managed to learn Portuguese. Today, I am still studying the effects of rainforest fragmentation, but on the rapidly vanishing forests of the Amazon.

As it turned out, the battle over north Queensland's rainforests was far from over in 1987. The Queensland government left no stone unturned in its efforts to defeat the World Heritage nomination. For example, Queensland tried to block the nomination through an injunction from Australia's High Court, partly based on its assertion that the area was not sufficiently unique biologically to warrant such protection. In its decision, the High Court likened Queensland's case to "Alice in Wonderland." Having lost that fight, the state tried other fronts. At the 1988 IUCN[5] General Assembly in Costa Rica, for example, I saw a Queensland delegation attempt—again unsuccessfully—to derail the nomination. Despite such dogged intransigence, the World Heritage listing was finalized in late 1988.

North Queensland is a different place today. Cairns and the other coastal cities have grown enormously, much of the growth fueled by tourism. The dire predictions of conservative politicians that World Heritage listing would cripple the region's economy were clearly unfounded. Millaa Millaa has several new shops and a second takeaway, though it is still a one-pub town. Perhaps the biggest change has been in the character of the populace; many of the fiercely independent loggers left the Atherton Tableland for the gold mines of the Queensland outback.

When studying the impact of humankind on tropical ecosystems, it is difficult not to feel pessimistic about the future. At present rates of deforestation, the equivalent of seventy football fields are being destroyed every minute, and much of the remainder fragmented or logged. But all is not lost. The bitter battle to protect north Queensland's rainforests is being fought again and again throughout the tropical world, and in some cases the conservationists are winning.

Below you will find information on organizations that are fighting to save the world's vanishing rainforests. I hope you'll join one of them and get involved in the battle. God knows, the forests need your help.

5. International Union for the Conservation of Nature and Natural Resources.

RAINFOREST CONSERVATION ORGANIZATIONS

Rainforest Action Network
221 Pine St., Suite 500
San Francisco, CA 94104
Phone: 415-398-4404
To join, email: ranmembers@ran.org
Web page: <http://www.ran.org>
Contact: Patrick Reinsborough,
email: rags@ran.org

Conservation International
2501 M Street, NW
Suite 200
Washington, D.C. 20037
Phone: 202-429-5660
Web address:
<http://www.conservation.org>

Amazon Watch
Contact: Atossa Soltani
Web address:
<http://www.amazonwatch.org>

Australian Rainforest Conservation Society, Inc.
19 Colorado Ave.
Bardon, Queensland 4065
Australia
Contact: Dr. Aila Keto, Dr. Keith Scott
Phone: 61-7-3368-1318
Fax: 61-7-3368-3938
Email: arcs@gil.com.au

To receive free, highly topical news about rainforest conservation and to link to a variety of related Web sites, access the following Web address: <http://www.forests.org>.

GLOSSARY

Bommie: A vertical column of tropical reef, comprising many types of coral and festooned with sponges, giant clams, and other marine species. Usually found in shallow water (<40 feet deep).

Climbing rattan: A spiny, highly modified palm (*Calamus* spp.) that is adapted for ascending into the canopy of rainforests, native to tropical Asia, New Guinea, and Australia. As they shake in the wind, many rattans use fishhook-like spines to fasten onto other plants and ratchet themselves up into the canopy. In heavily disturbed forests, rattans (called wait-a-whiles or lawyer vines in Australia) can form almost impenetrable masses. Most cane furniture is made from the stems of rattans.

Epiphyte: A highly specialized plant that grows while attached to tree limbs and trunks, often high in the canopy. Common in tropical forests. Because their roots do not touch the soil, epiphytes must extract nutrients from falling leaf litter (for this reason many are bowl-shaped to catch litter, and most grow slowly). In north Queensland, many orchids and ferns are epiphytic.

Liana: A woody vine, common in tropical forests. Lianas grow up tree trunks and produce large flushes of leaves in tree crowns. They compete with trees for nutrients, water, and light, and create structural stresses that can cause limb breakage or tree death. Some lianas grow to over two feet in diameter. There are many liana species; most are light-loving and thus are good indicators of past disturbances, such as cyclones or logging.

Niche: The combination of space and resources used or required by a species for survival. According to classical ecological theory, no two species can have exactly the same niche and coexist. Each must differ somehow in its use of space or resources, or competition between them is likely to be severe.

Olfaction: Sense of smell (i.e., ability to sense airborne chemicals).

AUSTRALIAN SPECIES MENTIONED IN THE TEXT

MONOTREMES

Short-beaked echidna *Tachyglossus aculeatus*
Platypus *Ornithorhynchus anatinus*

PREDATORY MARSUPIALS

Brown antechinus *Antechinus stuartii*
Yellow-footed antechinus *Antechinus flavipes*
Atherton antechinus *Antechinus godmani*
Tasmanian devil *Sacrophilus harrisii*
Spotted-tailed quoll *Dasyurus maculatus*
Long-nosed bandicoot *Perameles nasuta*
Thylacine (extinct) *Thylacinus cynocephalus*

HERBIVOROUS MARSUPIALS

Koala *Phascolarctos cinereus*
Long-tailed pygmy possum *Cercartetus caudatus*
Coppery brushtail possum *Trichosurus vulpecula johnstoni*
Green ringtail possum *Pseudocheirops archeri*
Herbert River ringtail possum *Pseudochirulus herbertensis*
Lemuroid ringtail possum *Hemibelideus lemuroides*
Musky rat-kangaroo *Hypsiprimnodon moschatus*
Unadorned rock-wallaby *Petrogale inornata*
Lumholtz's tree-kangaroo *Dendrolagus lumholtzi*

PLACENTAL MAMMALS

Fawn-footed melomys *Melomys cervinipes*
Cape York rat *Rattus leucopus*
Bush rat *Rattus fuscipes*
Giant white-tailed rat *Uromys caudimaculatus*
Water rat *Hydromys chrysogaster*
Spectacled flying fox *Pteropus conspicillatus*
Dingo *Canis familiaris dingo*
European rabbit (introduced) *Oryctolagus cuniculus*
Red fox (introduced) *Vulpes vulpes*
Feral cat (introduced) *Felis domesticus*
Feral pig (introduced) *Sus scrofa*

BIRDS

Southern cassowary *Casuarius casuarius*
Emu *Dromaius novahollandiae*
Australian brush turkey *Alectura lathami*
Boobook owl *Ninox novaeseelandiae*
Rufous owl *Ninox rufa*
Barn owl *Tyto alba*
Lesser sooty owl *Tyto multipunctata*
Crimson rosella *Platycercus elegans*
Laughing kookaburra *Dacelo novaeguineae*
Rainbow bee-eater *Merops orientalis*
Chowchilla *Orthonyx spaldingii*
Eastern whipbird *Psophodes olivaceus*
Spotted catbird *Ailuroedus melanotis*

FROGS AND REPTILES

Cane toad *Bufo marinus*
Crook car-horn *Cophixalus ornatus*
Boyd's forest dragon *Gonocephalus boydii*
New Zealand giant gecko *Hoplodactylus delcourti*
Northern death adder *Acathophis praelongus*
Red-bellied blacksnake *Pseudechis porphryiacus*
Taipan *Oxyuranus scutellatus*
Eastern brown snake *Pseudonaja textilis*
Brown treesnake *Boiga irregularis*
Carpet python *Morelia spilota*
Amesthytine python *Morelia amesthyna*
Kimberlies python *Morelia carinata*
Saltwater crocodile *Crocodilus porosus*

INVERTEBRATES

Brown rainforest leech *Chtonobdella* sp.
Tiger leech *Erpobdella* sp.
Australian paralysis tick *Ixodes holocyclus*
Australian redback spider *Latrodectus hasselti*
Greyback beetle *Dermolepida albahirtum*

PLANTS

Wait-a-while (lawyer vine, or climbing rattan) *Calamus* spp.
Stinging tree *Dendrocnides moroides*
Bird's-nest fern *Drynaria rigidula*
Banksia *Banksia* spp.
Protea *Protea* spp.
Queensland walnut *Bielschmedia bancrofti*
Gum tree *Eucalyptus* spp.
Wattle tree *Acacia* spp.
She-oak tree *Allocasuarina* spp.
Sasparilla tree *Alphitonia petriei*
Lantana (introduced) *Lantana camara*
Potato vine (introduced) *Solanum dallachi*
Tobacco tree (introduced) *Solanum mauritianum*
Caribbean pine (introduced) *Pinus caribbea*

SUGGESTED READINGS

BOOKS FOR GENERAL AUDIENCES

About Tropical Forests

Caufield, C. 1986. *In the Rainforest: Report from a Strange, Beautiful, Imperiled World.* University of Chicago Press, Chicago, Ill.

Forsyth, A., et al. 1987. *Tropical Nature.* Macmillan, New York.

Lambertini, M. 2000. *A Naturalist's Guide to the Tropics.* University of Chicago Press, Chicago, Ill.

Myers, N., and N. J. Myers. 1992. *The Primary Source: Tropical Forests and Our Future.* W. W. Norton and Co., New York.

About Australasia

Berra, T. M. 1998. *A Natural History of Australia.* Academic Press, New York.

Flannery, T. 1994. *The Future Eaters: An Ecological History of the Australasian Lands and People.* Reed New Holland, Sydney, Australia.

Flannery, T. 1998. *Throwim Way Leg: Tree-kangaroos, Possums, and Penis Gourds—On the Track of Unknown Mammals in Wildest New Guinea.* Atlantic Monthly Press, New York.

Hughes, R. 1987. *The Fatal Shore: The Epic of Australia's Founding.* Vintage Books, London.

Strahan, R., and P. Conder. 1998. *The Incomplete Book of Australian Mammals.* Seven Hills Book Distributors, Sydney, Australia.

BOOKS AND ARTICLES FOR ADVANCED READERS

About Tropical Forests

Laurance, W. F., and R. O. Bierregaard, Jr., eds. 1997. *Tropical Forest Remnants: Ecology, Management, and Conservation of Fragmented Communities.* University of Chicago Press, Chicago, Ill.

Terborgh, J. 1992. *Diversity and the Tropical Rain Forest.* W. H. Freeman and Co., San Francisco.

Whitmore, T. 1998. *An Introduction to Tropical Rain Forests.* Oxford University Press, Oxford, England.

About Australasia

Cogger, H. G. 1992. *Reptiles and Amphibians of Australia.* Cornell University Press, Ithaca, N.Y.

Ford, H. A. 1989. *Ecology of Birds: An Australian Perspective.* Surrey Beatty and Sons, Chipping Norton, New South Wales, Australia.

Hutton, D., et al. 1999. *History of the Australian Environmental Movement.* Cambridge University Press, Cambridge, England.

Lunney, D., ed. 1993. *Conservation of Australia's Forest Fauna.* Royal Zoological Society of New South Wales, Sydney, Australia.

Strahan, R., ed. 1996. *The Mammals of Australia.* Smithsonian Institution Press, Washington, D.C.

Selected Articles from the Rainforest Fragmentation Project

Laurance, W. F. 1990. Comparative responses of five arboreal marsupials to tropical forest fragmentation. *Journal of Mammalogy* 71:641–653.

Laurance, W. F. 1990. The top end of Down Under. *Wildlife Conservation Magazine,* July-August, 26–45.

Laurance, W. F. 1991. Ecological correlates of extinction proneness in Australian tropical rainforest mammals. *Conservation Biology* 5:79–89.

Laurance, W. F. 1991. Edge effects in tropical forest fragments: Application of a model for the design of nature reserves. *Biological Conservation* 57:205–219.

Laurance, W. F., and E. Yensen. 1991. Predicting the impacts of edge effects in fragmented habitats. *Biological Conservation* 55:77–92.

Laurance, W. F. 1994. Rainforest fragmentation and the structure of small mammal communities in tropical Queensland. *Biological Conservation* 69:23–32.

Laurance, W. F. 1995. Extinction and survival of rainforest mammals in a fragmented tropical landscape. In *Landscape Approaches in Mammalian Ecology and Conservation,* ed. W. Z. Lidicker, Jr., 46–63. University of Minnesota Press, Minneapolis.

Laurance, W. F. 1997. Responses of mammals to rainforest fragmentation in tropical Queensland: A review and synthesis. *Wildlife Research* 24:603–612.